Performance-Based Tests
with Answer Key

HOLT, RINEHART AND **WINSTON**

Harcourt Brace & Company

Austin • New York • Orlando • Atlanta • San Francisco • Boston • Dallas • Toronto • London

ISBN 0-03-051443-6   5   021   00

Contents

About Performance-Based Tests

Performance-Based Tests consists of blackline masters for eight activity-oriented unit tests. Teaching strategies and student guidelines precede each unit test. Each performance-based test can take up to two class periods. The time required and the level of difficulty for each test can be reduced by omitting some of the questions. Sample answers are provided at the end of this booklet to help you assess students' responses. A master materials list is also included in the answer key.

Unit 1: Studying the Earth

Performance-Based Test: Using Topographic Maps

Teacher's Notes

Overview

The students compare topographic maps drawn at different scales. The students are asked to gather information from different map elements, especially map scales and contour lines. The students then use the maps to solve a variety of problems.

Materials (per student or group)

ruler
protractor (or magnetic compass)
calculator
English-metric conversion table
pipe cleaner, or twist tie

Preparation

Assemble the materials prior to the test. Review topographic maps with the students if necessary.

Performance

At the end of the test, students should turn in the following items:

- completed maps
- completed test sheets

Evaluation

The following is a recommended breakdown for evaluating student performance.

10% appropriate use of materials and equipment
10% accuracy in map reading
15% ability to compare maps drawn at different scales
15% ability to infer topography from contour lines
25% identification of critical factors in problem solving
25% accuracy of answers to questions

M O D E R N   E A R T H   S C I E N C E

Unit 1: Studying the Earth
Performance-Based Test: Using Topographic Maps

Objectives
Demonstrate your ability to read topographic maps accurately. Use the maps provided to solve problems in a variety of hypothetical situations.

Background
Maps are designed to convey information. All map elements, including the legend and title, contain useful information. However, reading or interpreting a map accurately is a skill that takes practice. It requires being able to identify a map's symbols and being able to determine direction and distance.

Maps are designed for a variety of purposes. The choice of map elements, such as the scale and contour interval, depends largely on the purpose of the map. Comparing maps whose scales, contour intervals, and systems of measurement differ is challenging but often necessary. Because map designs can vary widely, knowing what type of map is best suited for a particular problem is also necessary.

Before You Begin
Read the following guidelines for completing this test.
- Keep in mind that your teacher will be observing and grading your in-class behavior as well as your written responses. In particular, your teacher will be noting your ability to follow the given procedures, your adherence to classroom or laboratory safety, and your methods and reasoning in solving the problems.
- Try not to let what others are doing influence your work. Remember that a problem often has several acceptable solutions.
- Do not talk to other students unless you are working in a group. Talk only to members of your group and try not to disturb other students.
- Use only the materials provided.

Procedure
Use the assembled materials and the topographic maps on pages 8–10 to answer all the questions as completely as you can. The maps provided are portions of USGS maps that have been modified for your use.

M O D E R N   E A R T H   S C I E N C E

Unit 1: Performance-Based Test

Questions

Answer the following questions in the space provided. Support your answers by explaining your reasoning and showing calculations, if any.

1. Which of the three maps represents the largest area?

   _____

   _____

   _____

   _____

2. Compare the areas of the cities of Comanche, Carlsbad, and Del Norte. Which city has the largest area?

   _____

   _____

   _____

   _____

3. Find the highest point on each map, and mark these points with an *A*. Which map contains the highest point?

   _____

   _____

   _____

   _____

4. If you were using a compass in Comanche, Texas, in what direction would the compass needle point with respect to true north?

   _____

   _____

   _____

   _____

5. Which map is the least distorted? Why?

   _____

   _____

   _____

   _____

   _____

M O D E R N   E A R T H   S C I E N C E

Unit 1: Performance-Based Test

6. Briefly describe the view you might see if you were looking east from Del Norte?

_____

_____

_____

_____

looking west?

_____

_____

_____

looking north?

_____

_____

_____

7. Locate the stream nearest the radio tower west of Carlsbad, at the southeastern edge of the Ocotillo Hills. In which direction does the stream flow? Explain your answer.

_____

_____

_____

_____

8. Fill in the missing information in the table below. (The contour intervals on both the Carlsbad map and the Durango map are shown in feet.)

Map name	Verbal scale	Map scale	Contour interval
Carlsbad			
Comanche	1 cm = 1 km		10 m
Durango		1:250,000	

HRW material copyrighted under notice appearing earlier in this work.

5

M O D E R N   E A R T H   S C I E N C E

Unit 1: Performance-Based Test

9. Maps with different scales are useful for different purposes. Choose which map would be the most useful for each of the following activities. Explain your choices.
   a. planning a five-day hiking trip
   b. deciding where to build a new industrial plant
   c. picking a site for a new housing development

   _____

   _____

   _____

   _____

   _____

   _____

   _____

   _____

   _____

10. Can you determine distances with a map based on the metric system if your ruler is based on the English system of measurement? Explain.

    _____

    _____

    _____

    _____

    _____

11. On the appropriate maps, locate the roads that connect the following places:
    • the western edge of Carlsbad and the buildings in the state park
    • the eastern edge of Comanche and the town of Blanket
    • Del Norte and Monte Vista

    If it costs $600,000 to widen 1 km of road, which road would cost the most to widen? Approximately how much would it cost? (Hint: Use a pipe cleaner or twist tie to measure the curved parts of the roads.)

    _____

    _____

    _____

    _____

    _____

    _____

    _____

M O D E R N   E A R T H   S C I E N C E

Unit 1: Performance-Based Test

12. Which of the roads listed in Question 11 is the steepest?

_____

_____

_____

_____

_____

_____

13. Find the radio tower just west of Carlsbad. Is this the best location for the radio tower? If not, select a new location for the tower and mark it with a point labeled *B*. Explain why your new location, if any, is better.

_____

_____

_____

_____

_____

_____

_____

_____

14. A pilot radios you and tells you that she has crashed her plane after traveling 52 km from Monte Vista, Colorado. She reports being on a heading of 270° according to her magnetic compass. Magnetic declination for the Durango map is 12° E. Place a point labeled *B* where you would expect to find the pilot.

M O D E R N   E A R T H   S C I E N C E

Unit 1: Performance-Based Test

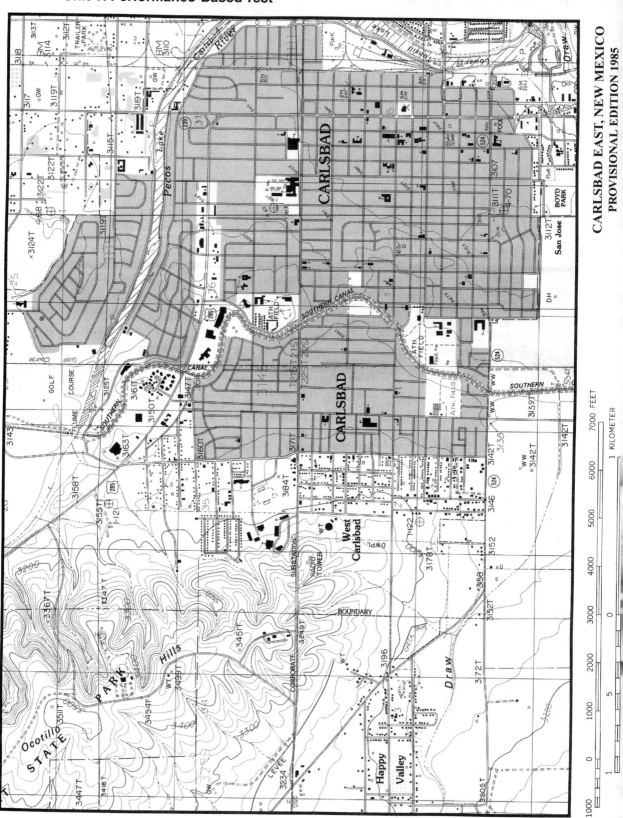

CARLSBAD EAST, NEW MEXICO
PROVISIONAL EDITION 1985

MODERN EARTH SCIENCE

Unit 1: Performance-Based Test

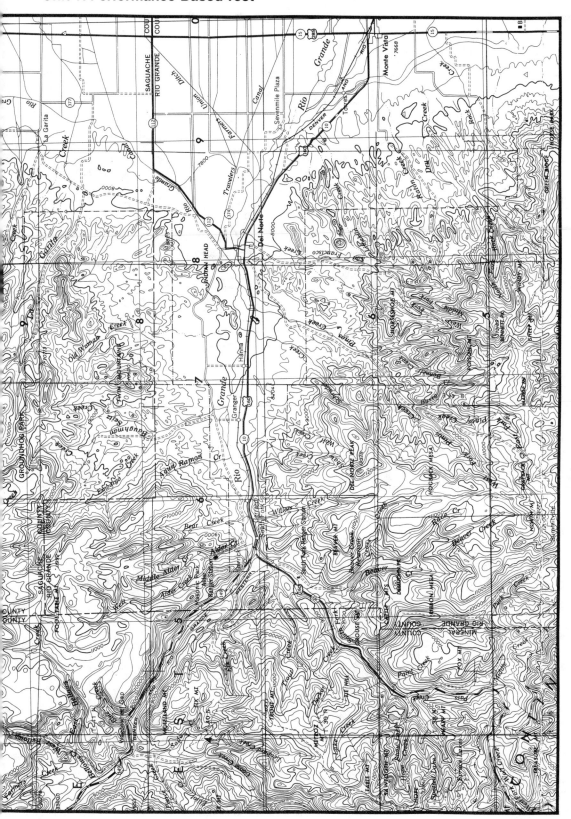

DURANGO, COLORADO
1945

Scale 1:250,000

M O D E R N   E A R T H   S C I E N C E

Unit 1: Performance-Based Test

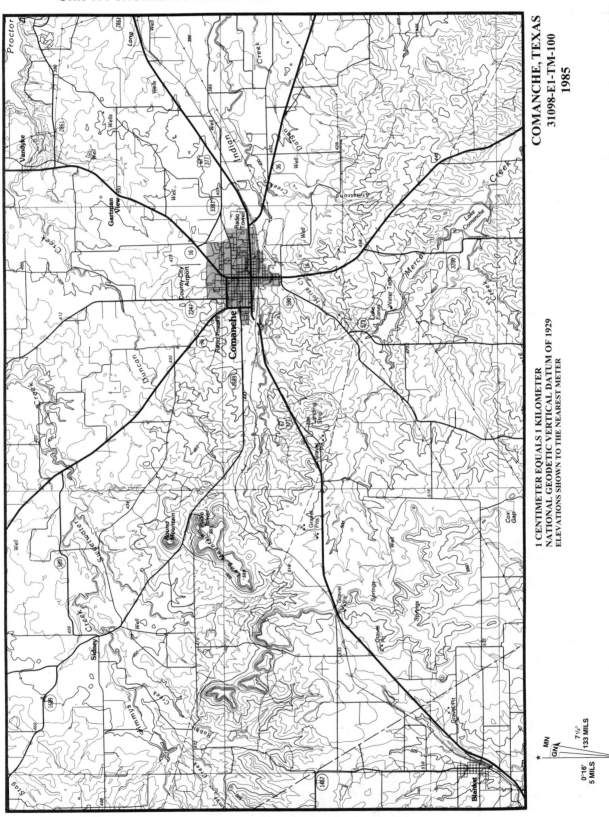

Unit 2: The Dynamic Earth
Performance-Based Test: Making Waves

Teacher's Notes

Overview
Students use a water tank to simulate the three main types of seismic waves. Students compare the motion of the different waves and their effect on rigid surface structures.

Materials (per student or group)
fish tank, or other large transparent container
3 fishing sinkers
3 cork stoppers
3 pushpins
string
grease pencil
strip of thin balsa wood or polystyrene
scissors
tape
paper towels

Preparation
Assemble the materials prior to the test. Review the safety procedures for spilled water and broken glass.

Safety Alert
Any water spilled on the floor should be mopped up immediately to prevent students from slipping. Broken tanks, pushpins, and scissors are sharp objects and should be handled carefully.

Performance
At the end of the test, students should turn in the following items:
- modified strip of balsa wood or polystyrene
- completed test sheets

Evaluation
The following is a recommended breakdown for evaluating student performance.
20% proper use of materials and equipment
30% quality of observations
25% identification of critical factors in problem solving
25% accuracy of answers to questions

M O D E R N   E A R T H   S C I E N C E

Unit 2: The Dynamic Earth
Performance-Based Test: Making Waves

Objectives
Use a water tank to try to create the three main types of seismic waves. Modify a rigid surface structure to make it less susceptible to earthquake damage.

Background
The energy released by earthquakes moves through the earth in the form of mechanical waves. Earthquakes generate three types of waves: compression, or P, waves; shear, or S, waves; and surface, or L, waves. Unlike electromagnetic waves, mechanical waves require a medium through which to travel. Differences in the medium greatly affect mechanical waves. By studying seismic waves, scientists can learn a great deal about the composition of the earth.

Before You Begin
Read the following guidelines for completing this test.
- Keep in mind that your teacher will be observing and grading your in-class behavior as well as your written responses. In particular, your teacher will be noting your ability to follow the given procedures, your adherence to classroom or laboratory safety, and your methods and reasoning in solving the problems.
- Try not to let what others are doing influence your work. Remember that a problem often has several acceptable solutions.
- Do not talk to other students unless you are working in a group. Talk only to members of your group and try not to disturb other students.
- Use only the materials provided.

Safety Alert
Any water spilled on the floor should be mopped up immediately to prevent slipping. Pushpins and scissors are sharp objects and should be handled carefully. Tanks should also be handled carefully so as not to break them.

Procedure
Step 1: Set up the tank, sinkers, and cork stoppers as shown in Figure 1 below. The first stopper should float on the surface and should have about 2 cm of extra string. The second stopper should be about two-thirds of the way from the bottom, and the third stopper should be about one-third of the way from the bottom. Be sure to draw the line on the side of the tank as shown to help you observe the motion of the corks.

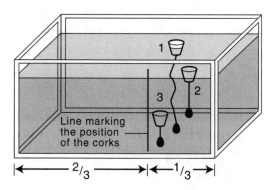

Figure 1

Unit 2: Performance-Based Test

Step 2: Make a loose fist near the bottom of the tank and point your thumb toward the stoppers (Figure 2). Quickly and firmly clench your fist without moving your arm. Record any motion of the three stoppers. You may have to place your hand closer to the stoppers to achieve any noticeable results.

_____

_____

_____

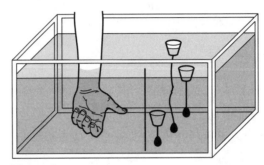

Figure 2

Step 3: Repeat Step 2, but with your thumb pointed toward the bottom of the tank. Record your observations.

_____

_____

Step 4: Open your fist and place your hand halfway into the water with your palm facing the stoppers (Figure 3). Move your hand from side to side. Record your observations.

_____

_____

_____

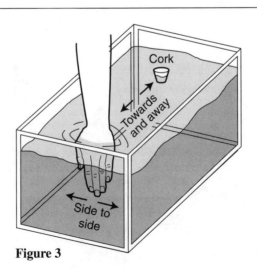

Figure 3

M O D E R N   E A R T H   S C I E N C E

Unit 2: Performance-Based Test

Step 5: Repeat Step 4, but this time move your hand toward and away from the corks. You should generate several waves at a time. Record your observations.

_____

_____

_____

Step 6: Empty the tank halfway and float the strip of balsa wood or polystyrene on the water (Figure 4). Make waves as in Step 5. Make the waves as large and as rapidly as you can without spilling the water. Observe the strip from the side. Record your observations, noting especially the motion at the ends and the middle of the strip.

_____

_____

_____

_____

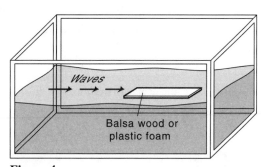

Figure 4

Questions

Answer the following questions in the space provided. Support your answers by explaining your reasoning.

1. Which type of wave did you create in Step 2? Explain.

_____

_____

_____

_____

2. Which type of wave did you create in Step 3? Explain.

_____

_____

_____

_____

M O D E R N   E A R T H   S C I E N C E

Unit 2: Performance-Based Test

3. Which type of wave did you try to create in Step 4? Explain your observations.

_____

_____

_____

_____

4. Which type of wave did you create in Step 5? Explain.

_____

_____

_____

_____

5. Based on the motion of the surface stopper, how do you think the particles of the medium move to create surface waves?

_____

_____

_____

_____

6. If a seismograph doesn't record any shear waves from an earthquake, what can you infer about the medium through which the seismic waves passed?

_____

_____

_____

_____

7. From your observations of the floating strip in Step 6, explain why surface waves do so much damage to buildings that cover large areas.

_____

_____

_____

_____

8. Suppose that the floating strip is the foundation of a building. Use the assembled materials to make the strip better able to withstand an earthquake. Briefly explain your modifications.

_____

_____

_____

_____

Unit 3: Composition of the Earth
Performance-Based Test:
Prospecting for Ores and Oil

Teacher's Notes

Overview
Students use their knowledge of the properties and origins of ores to select possible mining sites. Students then use graphs of oil-well production to identify profitable wells.

Materials (per student or group)
clear plastic container with lid
a mixture of silt, iron filings (or a dark, crushed, dense mineral such as galena or pyrite), and light-colored sand

Preparation
Assemble the materials prior to the test. You may wish to split the class into two groups, one for ores and one for petroleum. You may wish to inform the class that different people will have different tasks.

Performance
At the end of the test, students should turn in the following items:
• completed test sheets

Evaluation
The following is a recommended breakdown for evaluating student performance.
15% appropriate use of materials and equipment
30% quality of observations
30% identification of critical factors in problem solving
25% accuracy of answers to questions

M O D E R N   E A R T H   S C I E N C E

Unit 3: Composition of the Earth
Performance-Based Test:
Prospecting for Ores and Oil

Objectives
Use your knowledge of the physical and chemical properties of ores to select possible mining sites. Analyze graphs of oil-well production to identify profitable wells.

Background
Today's societies are increasingly dependent on sources of ore minerals. The search for earth's minerals is no longer the simple one- or two-person operation of early prospectors. It is highly organized and involves the latest scientific methods and equipment. However, as with early prospecting, a knowledge of the properties of ore minerals is required for success.

Exploring for fossil fuels has also advanced as demand has increased. New methods have been developed not only to find these deposits but also to gauge potential production. One method of monitoring an oil well's productivity is to plot the history of the well's oil production. These histories indicate when pumping a well is no longer economically viable and how much oil remains in the well.

Before You Begin
Read the following guidelines for completing this test.
- Keep in mind that your teacher will be observing and grading your in-class behavior as well as your written responses. In particular, your teacher will be noting your ability to follow the given procedures, your adherence to classroom or laboratory safety, and your methods and reasoning in solving the problems.
- Try not to let what others are doing influence your work. Remember that a problem often has several acceptable solutions.
- Do not talk to other students unless you are working in a group. Talk only to members of your group and try not to disturb other students.
- Use only the materials provided.

Procedure
Fill the clear container about one-quarter full with a mixture of sand, silt, and particles of heavy metal, such as iron filings or crushed galena. Add water until the container is almost full. Swirl the contents of the container to thoroughly mix the solids and water. Observe the mixture as you continue to agitate the container. Gradually slow down until you are causing the liquid to barely move.

M O D E R N   E A R T H   S C I E N C E

Unit 3: Performance-Based Test

Questions

Answer the following questions in the space provided. Support your answers by explaining your reasoning.

Section I: The Location of Mineral Ores

1. What did you observe about the order in which the different materials settled to the bottom of the container?

_____

_____

_____

_____

2. How do you think the speed of the water as you swirled and agitated the jar affected the order in which the different materials settled?

_____

_____

_____

_____

3. How might the speed of the water affect heavy minerals, such as gold, silver, or tin, as they are washed down a stream or onto a beach?

_____

_____

_____

_____

_____

_____

4. On the map on page 24, which letters represent locations where placer gold deposits might be found? Why do you think there may be gold there? (Hint: There must be a source of gold in order for a placer deposit to form.)

_____

_____

_____

_____

_____

_____

_____

M O D E R N E A R T H S C I E N C E

Unit 3: Performance-Based Test

5. Which letters on the map represent locations where you might find minerals formed through contact metamorphism? Explain.

_____

_____

_____

_____

_____

_____

_____

6. While still in its original rock, gold is often associated with sulfide minerals. Using the reported findings on the map, pick the best locations to prospect for this type of deposit. Explain your choices.

_____

_____

_____

_____

_____

_____

_____

_____

Section II: The Production Rates of Oil Wells

The graphs on pages 25–26 plot the production histories of six wells. Refer to these graphs to answer the remaining questions. Remember to justify all of your answers.

7. Which well has produced the most oil to date?

_____

_____

_____

_____

8. Which well is producing at the highest rate?

_____

_____

_____

_____

_____

M O D E R N   E A R T H   S C I E N C E

Unit 3: Performance-Based Test

9. Which well is producing at the lowest rate?

_____

_____

_____

_____

10. Which well will probably be the first to be abandoned as unprofitable?

_____

_____

_____

_____

11. Which well probably has the most oil left to produce?

_____

_____

_____

_____

12. Suppose that you are an oil company's expert in buying existing oil wells. The six wells whose production rates are graphed on pages 25–26 are up for sale. List your buying preferences in order from best to worst. For each well, explain its ranking.

Well name: _____

_____

_____

_____

_____

_____

_____

Well name: _____

_____

_____

_____

_____

_____

_____

M O D E R N   E A R T H   S C I E N C E

Unit 3: Performance-Based Test

Well name: _____

_____

_____

_____

_____

_____

_____

Well name: _____

_____

_____

_____

_____

_____

_____

Well name: _____

_____

_____

_____

_____

_____

_____

Well name: _____

_____

_____

_____

_____

_____

_____

RW material copyrighted under notice appearing earlier in this work.

23

Unit 3: Performance-Based Test

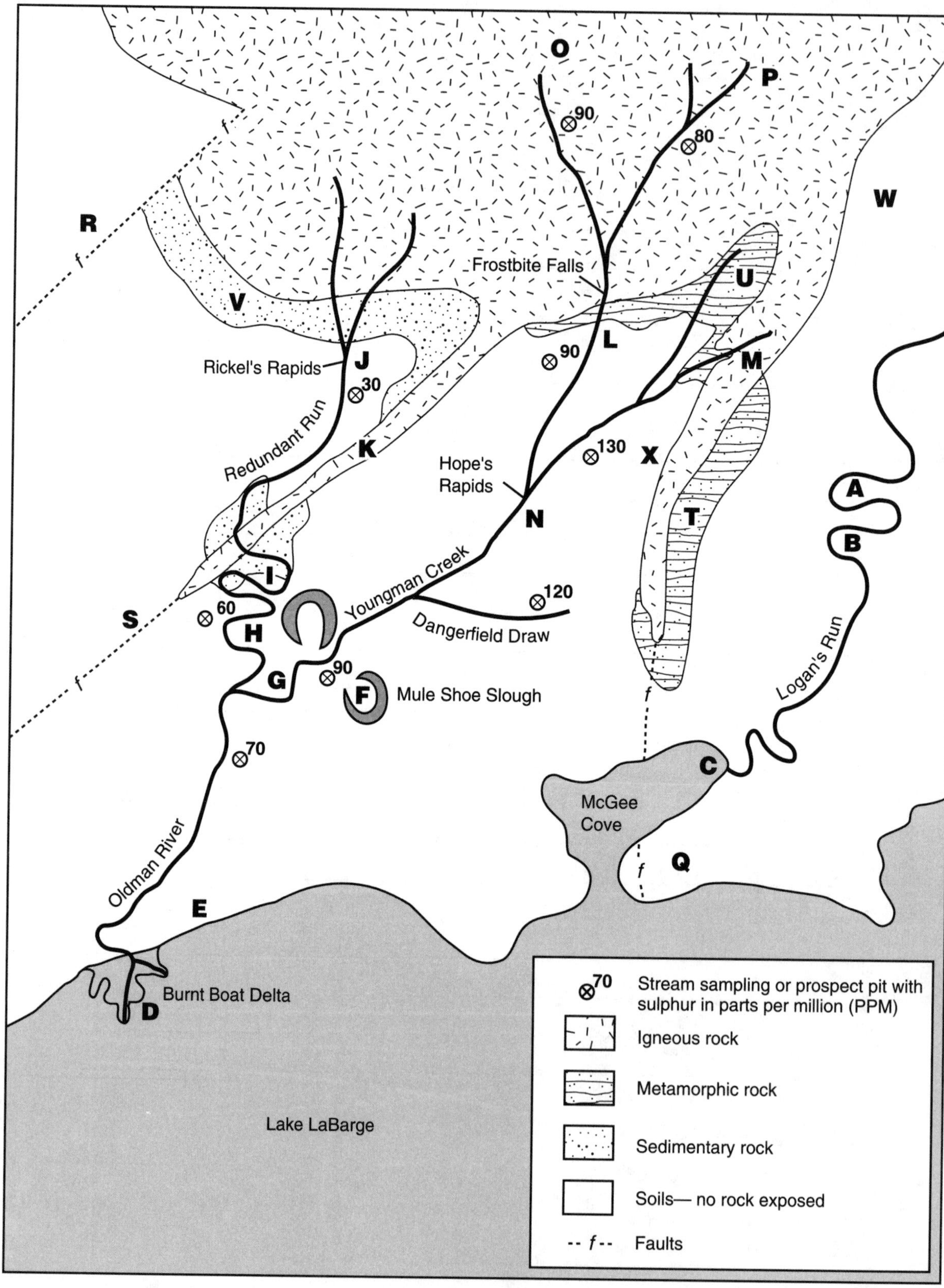
O
P
90
80
W
R
Frostbite Falls
U
V
L
90
M
Rickel's Rapids
J
30
Redundant Run
K
130
X
Hope's
Rapids
A
N
T
B
I
Youngman Creek
120
60
S
H
Dangerfield Draw
90
G
F
Mule Shoe Slough
Logan's Run
f
C
70
McGee
Cove
Oldman River
Q
E
Burnt Boat Delta
D
Lake LaBarge
70
Stream sampling or prospect pit with sulphur in parts per million (PPM)
Igneous rock
Metamorphic rock
Sedimentary rock
Soils— no rock exposed
-- f -- Faults

Unit 3: Performance-Based Test

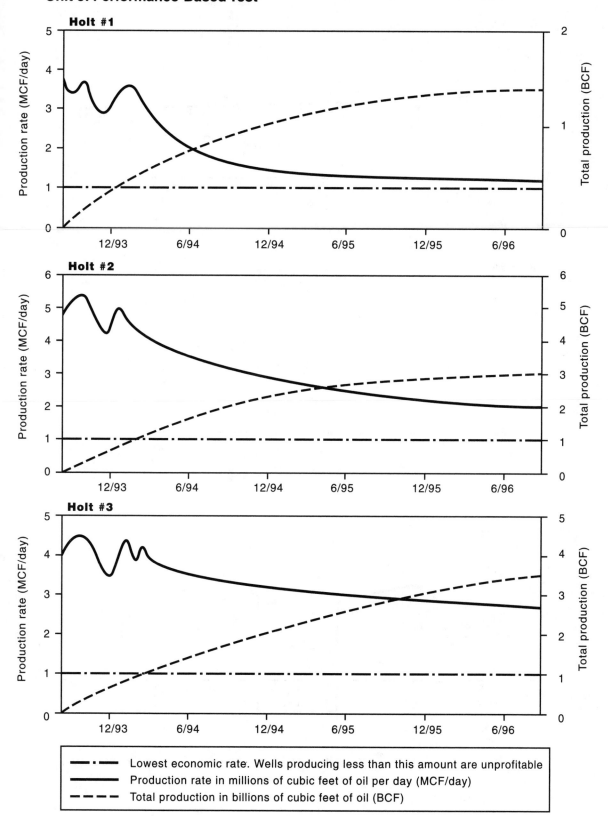

Holt #1

Holt #2

Holt #3

— · — · — Lowest economic rate. Wells producing less than this amount are unprofitable

———— Production rate in millions of cubic feet of oil per day (MCF/day)

— — — — Total production in billions of cubic feet of oil (BCF)

MODERN EARTH SCIENCE

Unit 3: Performance-Based Test

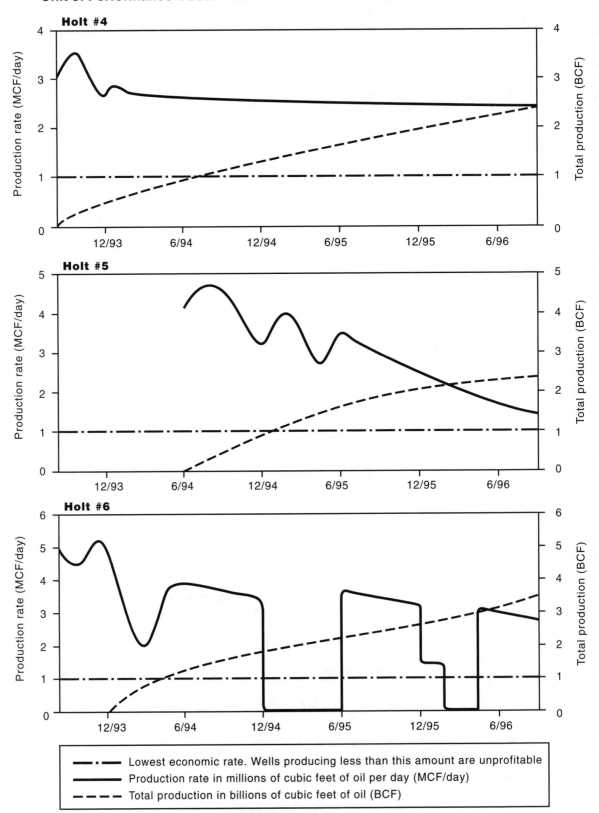

Legend:

—·—·— Lowest economic rate. Wells producing less than this amount are unprofitable

——— Production rate in millions of cubic feet of oil per day (MCF/day)

- - - - Total production in billions of cubic feet of oil (BCF)

Unit 4: Reshaping the Crust
Performance-Based Test: Controlling Erosion

Teacher's Notes

Overview
Students use a stream table to investigate soil conservation practices and the factors affecting rates of erosion. Students apply their results to hypothetical problems in soil conservation.

Materials (per student or group)
stream table, or a setup similar to that on page 248 of the textbook
watering can, or spray bottle
ruler
grease pencil
plastic fork
tongue depressors, or similar wooden splints
clock or watch with a second hand
sheet of plastic window screening large enough to cover the sand in the setup
small beaker

Preparation
Assemble the materials prior to the test. Have students read through the activity to ensure quick completion of the hands-on part of the test.

Safety Alert
Water spills must be mopped up immediately to prevent students from slipping.

Performance
At the end of the test, students should turn in the following items:
- completed data table
- completed test sheets

Evaluation
The following is a recommended breakdown for evaluating student performance.
15% appropriate use of materials and equipment
30% quality of observations
30% identification of critical factors in problem solving
25% accuracy of answers to questions

M O D E R N   E A R T H   S C I E N C E

Unit 4: Reshaping the Crust

Performance-Based Test: Controlling Erosion

Objectives

Test several methods of soil conservation and evaluate their effectiveness. Use your results to solve problems in soil conservation.

Background

One of our most important natural resources is high-quality topsoil. Without this resource, we could not grow food. As valuable as this resource is, it gets very little public attention. While it is true that farmers and ranchers share the major responsibility for protecting this resource, we all suffer from the results of soil erosion.

Millions of dollars are spent annually dredging eroded soil from rivers and bays. This eroded soil not only threatens our use of waterways but also decreases the available wetlands and water sources used by wildlife. Even the soil eroding from our lawns may cause problems by clogging storm drains and causing flooding.

Before You Begin

Read the following guidelines for completing this test.

- Keep in mind that your teacher will be observing and grading your in-class behavior as well as your written responses. In particular, your teacher will be noting your ability to follow the given procedures, your adherence to classroom or laboratory safety, and your methods and reasoning in solving the problems.
- Try not to let what others are doing influence your work. Remember that a problem often has several acceptable solutions.
- Do not talk to other students unless you are working in a group. Talk only to members of your group and try not to disturb other students.
- Use only the materials provided.

Safety Alert

Water spills must be mopped up immediately to prevent slipping.

Procedure

Step 1: Create a smooth slope of moist sand in a stream table. With a grease pencil, trace the angle of the slope on the side of the stream table. Water the upper part of the slope evenly until a stream forms. Record the height from which you pour the water, the amount of water you use, and the time it takes to form a stream. You may need several containers of water to obtain the desired effect.

Step 2: Use a small beaker to measure the amount of sand that eroded. Record this amount in the table on the next page.

Step 3: Restore the slope to its original shape using the line you marked with the grease pencil as a reference. Use a fork to plow furrows perpendicular to the angle of the slope. Create a deeper furrow every so often, as shown in Figure 1 on the next page. Add water to the slope exactly as you did in Step 1, and record the amount of eroded sand in the data table.

Unit 4: Performance-Based Test

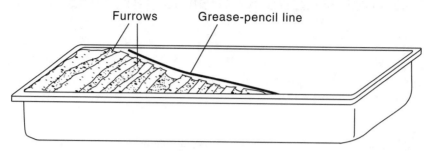

Furrows   Grease-pencil line

Figure 1

Step 4: Restore the slope. Place a plastic screen on the slope. Water the slope, and record the amount of eroded sand.

Step 5: Using the diagram below as a guide, use tongue depressors to create a series of terraces on the slope. Make sure you align the tops of the tongue depressors with the line you marked with the grease pencil. Also be sure that the tongue depressors are slightly higher than the sand behind them. Water the slope, and record the amount of eroded sand.

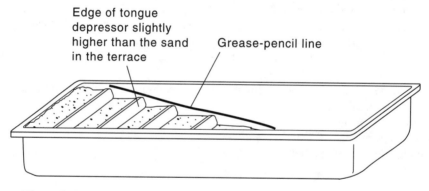

Edge of tongue depressor slightly higher than the sand in the terrace   Grease-pencil line

Figure 2

Step 6: If time allows, repeat some or all of the setups at steeper angles.

Data Table

Conservation method	Amount of eroded sand

M O D E R N   E A R T H   S C I E N C E

Unit 4: Performance-Based Test

Questions

Answer the following questions in the space provided. Support your answers by explaining your reasoning.

1. What method of soil conservation does the screen represent?

_____

_____

2. Why was it important to water the slope the same way for each method of soil conservation?

_____

_____

_____

3. According to your results, which method of soil conservation protected the soil the best? Why do you think this method was the most effective?

_____

_____

_____

_____

_____

_____

4. How would you expect your results to be different if the angle of the slope were steeper? Explain.

_____

_____

_____

_____

5. Which conservation method would you recommend for gently sloping hills?

_____

_____

_____

_____

M O D E R N   E A R T H   S C I E N C E

Unit 4: Performance-Based Test

6. Which conservation method would you recommend for steeply sloping hills?

_____

_____

_____

_____

7. Try to think of what objections a farmer might have to each of the following methods of soil conservation.

contour plowing

_____

_____

_____

_____

plant cover

_____

_____

_____

_____

terracing

_____

_____

_____

_____

8. Which method(s) would also work against wind erosion? Explain your choice(s).

_____

_____

_____

_____

_____

_____

Unit 5: The History of the Earth
Performance-Based Test: Writing Geologic History

Teacher's Notes

Overview

Students use geologic columns from several areas to construct regional columns. They make inferences about the region's geologic history by comparing the regional and local columns.

Materials (per student or group)

ruler
colored pencils

Preparation

Assemble the materials prior to the test. You may wish to review geologic columns with the class prior to the test.

Performance

At the end of the test, students should turn in the following items:
- 2 regional geologic columns
- regional map
- completed test sheets

Evaluation

The following is a recommended breakdown for evaluating student performance.
10% appropriate use of materials and equipment
20% presentation of geologic columns and map
20% ability to interpret geologic columns and make inferences based on the columns
25% identification of critical factors in problem solving
25% accuracy of answers to questions

M O D E R N   E A R T H   S C I E N C E

Unit 5: The History of the Earth

Performance-Based Test: Writing Geologic History

Objectives
Use geologic columns from several areas to construct regional geologic columns. Make inferences about the region's geologic history by comparing the regional and local columns.

Background
Much of the history of how the earth's landscapes and environments have changed lies in the rock record. However, no single area on the earth contains a record of all geologic time. To compile a complete history, geologists use clues such as fossils, rock type, and mineral content to combine local geologic columns into larger regional ones.

Once a regional geologic column is assembled, geologists try to decipher the meaning of gaps or changes in rock type across the region. They also try to reconstruct the landscape and environment of the region at different times. In doing so, geologists gain an understanding of the processes responsible for the earth's complex geologic history.

Before You Begin
Read the following guidelines for completing this test.
- Keep in mind that your teacher will be observing and grading your in-class behavior as well as your written responses. In particular, your teacher will be noting your ability to follow the given procedures, your adherence to classroom or laboratory safety, and your methods and reasoning in solving the problems.
- Try not to let what others are doing influence your work. Remember that a problem often has several acceptable solutions.
- Do not talk to other students unless you are working in a group. Talk only to members of your group and try not to disturb other students.
- Use only the materials provided.

Procedure
Step 1: In Figure 1 below, note how a geologist might use a geologic column to record the different sediments present in a shoreline.

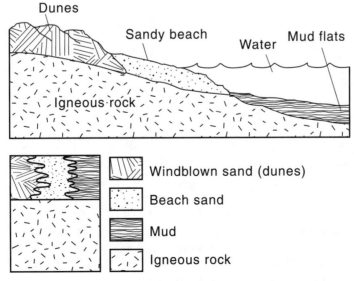

Figure 1. A modern beach (top) and the beach represented in a geologic column (bottom)

M O D E R N   E A R T H   S C I E N C E

Unit 5: Performance-Based Test

Step 2: Use the geologic columns on page 39 to construct two regional geologic columns in the blank columns below. Notice that the geologic columns on page 39 are divided into a western and an eastern group. Create one regional geologic column for each group of three, local, geologic columns. The thickness of a rock layer in a regional column should equal the maximum thickness attained by that rock layer in any of the three corresponding columns.

West

East

Step 3: The geologic columns on page 39 correspond to the locations labeled on the base map on page 40. On the base map, briefly sketch the surface features of the landscape suggested by the oldest sedimentary layer. Be sure to label each feature and include shorelines to distinguish aquatic environments from terrestrial environments.

M O D E R N   E A R T H   S C I E N C E

Unit 5: Performance-Based Test

Questions
Answer the following questions in the space provided. Support your answers by explaining your reasoning.

1. How do you explain the distribution of the geologic columns on the base map?

   _____

   _____

   _____

   _____

2. What does the wavy surface of the lowest rock layer in this region represent?

   _____

   _____

   _____

   _____

3. What was the source of the sediments that make up the lowest sedimentary layer in the geologic columns? Justify your answer.

   _____

   _____

   _____

   _____

   _____

4. Suggest an index fossil for the geologic columns of this region.

   _____

   _____

   _____

   _____

5. Explain the existence of a basalt layer in all of the geologic columns.

   _____

   _____

   _____

   _____

M O D E R N   E A R T H   S C I E N C E

Unit 5: Performance-Based Test

6. What type of geological feature is evident in geologic column *C?* What is its relative age?

_____

_____

_____

_____

7. What type of igneous rock formation is evident in the upper part of geologic columns *C* and *D?* What is its relative age?

_____

_____

_____

_____

_____

8. What type of tectonic-plate interaction is suggested by the top layer in column *E?*

_____

_____

_____

9. Summarize the geologic history of the region, and cite specific evidence in the geologic record to support your interpretation.

_____

_____

_____

_____

_____

_____

_____

_____

_____

_____

_____

_____

Unit 5: Performance-Based Test

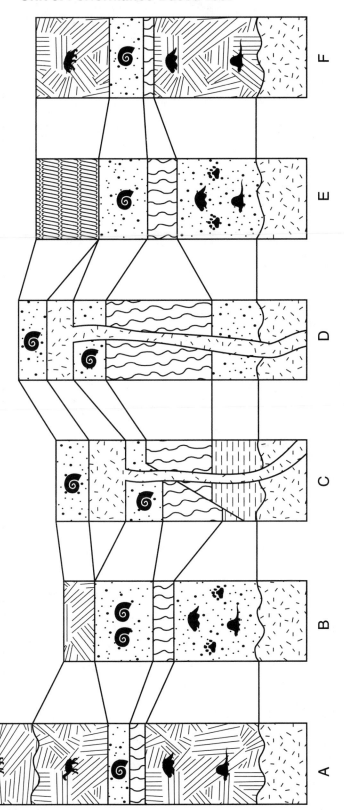

East

West

Fossil types		Rock types	
🐚	Snail		Wind-blown sand (dunes)
🐻	Type 1		Basalt flows or ash
🐗	Type 2		Deep water flysch (trench)
🦫	Type 3		Sandstone
🐟	Type 4		Shale
🐾	Tracks		Granite

M O D E R N   E A R T H   S C I E N C E

Unit 5: Performance-Based Test

Base Map

F ●

E ●

D ●

C ●

B ●

A ●

Unit 6: Oceans

Performance-Based Test:
Mapping the Ocean Floor

Teacher's Notes

Overview
Students use bathymetric data to draw a topographic map. Students then use the map to identify underwater features and to relate the features to ocean environments and plate tectonics.

Materials (per student or group)
soft-lead pencil (No. 2 or softer)
art gum, or soap eraser

Preparation
Assemble the materials prior to the test. Be sure students are familiar with contouring simple data sets. Warn students that contouring can involve quite a bit of erasing. Instruct students to draw lightly at first. You may wish to have extra copies of the base map on page 46 available.

Performance
At the end of the test, students should turn in the following items:
- completed topographic map
- completed test sheets

Evaluation
The following is a recommended breakdown for evaluating student performance.
10% appropriate use of materials and equipment
20% presentation of topographic map
20% accuracy of map
25% identification of critical factors in problem solving and feature identification
25% accuracy of answers to questions

M O D E R N   E A R T H   S C I E N C E

Unit 6: Oceans

Performance-Based Test:
Mapping the Ocean Floor

Objectives
Use data on water depths to draw a topographic map of the ocean floor. Use your completed map to identify underwater features, and relate the features to ocean environments and plate tectonics.

Background
Humankind has long sought to map the sea floor. The first attempts focused on shallow areas and shorelines to prevent ships from running aground and sinking. These early maps were made with a common tool: a plumb bob, or a weight on a string.

Today, depth measurements are largely made with *sonar,* an acronym for **so**und **n**avigation **a**nd **r**anging. Sonar relies on the travel time of a sound signal to determine distance, or range. The technique involves transmitting a sound signal underwater and timing how long it takes the signal to return after being reflected by the ocean floor. After adjusting the speed of sound for factors such as water temperature and density, the time is converted to a depth. Almost all naval vessels and some civilian ships operate sonar equipment continuously and report their data to the Navy.

Before You Begin
Read the following guidelines for completing this test.
- Keep in mind that your teacher will be observing and grading your in-class behavior as well as your written responses. In particular, your teacher will be noting your ability to follow the given procedures, your adherence to classroom or laboratory safety, and your methods and reasoning in solving the problems.
- Try not to let what others are doing influence your work. Remember that a problem often has several acceptable solutions.
- Do not talk to other students unless you are working in a group. Talk only to members of your group and try not to disturb other students.
- Use only the materials provided.

Procedure
Examine the sonar data on the base map on page 46. Select an appropriate contour interval, and contour the map as quickly and as cleanly as you can. Answer Question 1 to convert some of the data from seconds to meters before you begin. Make your lines very faint at first because you will probably be erasing often. Do not darken any lines until you are sure that you are satisfied with their placement. Remember to label the contour lines.

M O D E R N   E A R T H   S C I E N C E

Unit 6: Performance-Based Test

Questions

Answer the following questions in the space provided. Support your answers by explaining your reasoning.

1. Use the graph in the upper right corner of the map on page 46 to determine the ocean depth at points *H, I,* and *J.*

   _____

   _____

2. In the graph on page 46, why does the actual curve differ from the ideal curve?

   _____

   _____

   _____

   _____

3. Explain your choice of contour interval for the map.

   _____

   _____

   _____

   _____

4. At which of the locations *A–G* would you expect to find the coldest water?

   _____

   _____

   _____

   _____

5. At which of the locations *A–G* would you expect to find water with low salinity?

   _____

   _____

   _____

   _____

6. What type of ocean-floor feature is located at each of the lettered locations *A–G?*

   _____

   _____

   _____

   _____

M O D E R N   E A R T H   S C I E N C E

Unit 6: Performance-Based Test

7. The data are not evenly spaced around the map. How do you explain the distribution of the data?

_____

_____

_____

_____

_____

_____

8. At which of the locations *A–G* would the fishing for bottom-dwelling organisms be best? Why?

_____

_____

_____

_____

_____

9. What does the feature at location *E* suggest about tectonic activity in the area?

_____

_____

_____

_____

10. What do you see on the map that would support a change in sea level over time? Explain your answer.

_____

_____

_____

_____

_____

_____

11. If you were to select an area on the base map for more-detailed depth readings, where would it be? Explain your choice.

_____

_____

_____

_____

_____

_____

M O D E R N   E A R T H   S C I E N C E

M O D E R N   E A R T H   S C I E N C E

Unit 6: Performance-Based Test

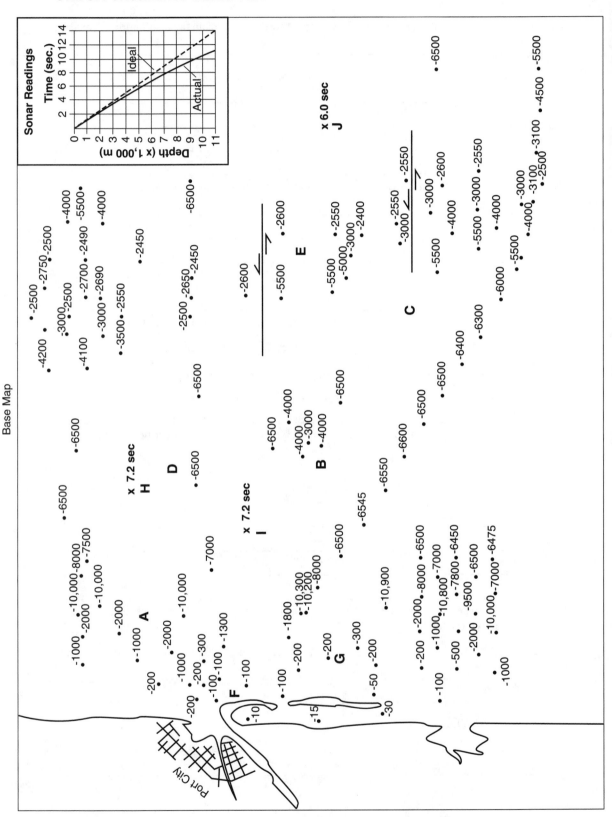

Base Map

Unit 7: Atmospheric Forces

Performance-Based Test: Creating Clouds

Teacher's Notes

Overview
Students create clouds in a bottle and identify the conditions required for cloud formation. Students then compare their experimental conditions to conditions in the atmosphere.

Materials (per activity station)
2–3 L clear plastic bottle with cap
matches
warm, cool, and cold water

Preparation
Assemble the materials prior to the test. Review the proper use of matches and fire-safety guidelines.

Safety Alert
Matches are a fire hazard. Be sure your students are familiar with the safety guidelines for working around open flames. Any water spilled on the floor should be mopped up immediately to prevent students from slipping. Loosely closed bottles may launch their caps when squeezed. Require eye protection.

Performance
At the end of the test, students should turn in the following items:
• completed test sheets

Evaluation
The following is a recommended breakdown for evaluating student performance.
15% appropriate use of materials and equipment
15% quality of observations
20% ability to identify and limit experimental variables
25% identification of critical factors in problem solving.
25% accuracy of answers to questions

M O D E R N   E A R T H   S C I E N C E

Unit 7: Atmospheric Forces

Performance-Based Test: Creating Clouds

Objectives

Create clouds in a bottle to identify the conditions required for cloud formation. Compare your experimental conditions to conditions in the atmosphere.

Background

Clouds are a common sight. The rain they bring can range from a welcome relief to a minor inconvenience to a natural disaster. Because we are dependent on and often at the mercy of rain, humankind has long tried to control clouds and rain but without much success. A key to control is our understanding of the natural processes involved. Although our understanding of how clouds and rain form is not complete, we now know enough to make some crude attempts at controlling clouds and rain.

Before You Begin

Read the following guidelines for completing this test.

- Keep in mind that your teacher will be observing and grading your in-class behavior as well as your written responses. In particular, your teacher will be noting your ability to follow the given procedures, your adherence to classroom or laboratory safety, and your methods and reasoning in solving the problems.
- Try not to let what others are doing influence your work. Remember that a problem often has several acceptable solutions.
- Do not talk to other students unless you are working in a group. Talk only to members of your group and try not to disturb other students.
- Use only the materials provided.

Safety Alert

Matches are a fire hazard. Be sure you are familiar with the safety guidelines for working around open flames. Any water spilled on the floor should be mopped up immediately to prevent slipping. Loosely closed bottles may launch their caps when squeezed. Wear safety goggles.

Procedure

Step 1: Fill the bottle about half full with water at room temperature. Drop a lit match into the bottle, and screw the cap on tightly.

Step 2: Make sure the top of the bottle is not pointed at anyone, including yourself. Squeeze the sides of the bottle as hard as you can and quickly release it. Do this several times until a cloud forms. Record the number of squeezes in the table on the next page. Empty the bottle, and squeeze it several times to clear the smoke from the bottle.

Step 3: Repeat Steps 1 and 2 with warm water, and record your results.

Step 4: Repeat Steps 1 and 2 with cold water, and record your results.

Step 5: Repeat Steps 1–4 but do not use any matches. Record your results.

M O D E R N   E A R T H   S C I E N C E

Unit 7: Performance-Based Test

Data Table

Water temperature	Match (yes or no)	Number of squeezes
Room temperature		
Warm		
Cold		
Room temperature		
Warm		
Cold		

Questions

Answer the following questions in the space provided. Support your answers by explaining your reasoning.

1. How did the water temperature affect the number of squeezes needed to create a cloud? Explain.

_____

_____

_____

_____

2. How did using matches affect the number of squeezes needed to create a cloud? Explain.

_____

_____

_____

_____

3. In some cases, small silver iodide crystals are dropped into clouds to seed them and make rain. What do you think the crystals do?

_____

_____

_____

_____

4. Based on your results, does the water temperature or the presence of smoke have a greater effect on cloud formation?

_____

_____

_____

_____

M O D E R N   E A R T H   S C I E N C E

Unit 7: Performance-Based Test

5. Write a hypothesis that explains how squeezing the bottle caused clouds to form.

_____

_____

_____

_____

_____

_____

_____

_____

6. Explain your results in terms of relative humidity and dew point.

_____

_____

_____

_____

_____

7. Estimate the probable amount of cloudiness and rainfall for the following climates. Rank the climates from most cloudy/rainy (1) to least cloudy/rainy (6).

_____ cool, moist, not dusty   _____ hot, dry, dusty

_____ cool, dry, dusty   _____ warm, moist, not dusty

_____ hot, moist, dusty   _____ warm, moist, dusty

Explain your ranking based on the results of your experiment.

_____

_____

_____

_____

_____

_____

_____

_____

_____

M O D E R N   E A R T H   S C I E N C E

M O D E R N   E A R T H   S C I E N C E

Unit 7: Performance-Based Test

8. In which of the climates listed in Question 7 do you think cloud seeding would be most effective? least effective?

_____

_____

_____

_____

_____

9. What might be done in the activity to improve the accuracy of data and the conclusions taken from the data?

_____

_____

_____

_____

_____

_____

_____

Unit 8: Studying Space

Performance-Based Test:
Putting Planets in Motion

Teacher's Notes

Overview
Students simulate planetary motion to verify Kepler's laws. They then apply these laws to solve hypothetical problems.

Materials (per student or group)
thread spool
stopwatch
small weights, such as washers
string
spring scale
scissors
metric ruler

Preparation
Assemble the materials prior to the test. Review the procedure and safety guidelines with the students. Students should work in groups of three or more. You may wish to assign groups as part of the preparation.

Safety Alert
Scissors are sharp objects and should be handled carefully. Also, make sure the string is stout and in good shape to prevent the weights from tearing the string. Use small weights, such as several large washers, and make sure that student groups are spaced far apart as they complete Steps 1–5. Require eye protection.

Performance
At the end of the test, students should turn in the following items:
- completed data table
- graph
- completed test sheets

Evaluation
The following is a recommended breakdown for evaluating student performance.
15% appropriate use of materials and equipment
15% clarity of data presentation
20% quality of observations
25% identification of critical factors in problem solving
25% accuracy of answers to questions

Unit 8: Studying Space

Performance-Based Test:
Putting Planets in Motion

Objectives

Investigate planetary motion by testing Kepler's laws. Use Kepler's laws to solve several hypothetical problems.

Background

In 1600, a young German named Johannes Kepler arrived in Prague. He quickly became the protégé of Tycho Brahe, the leading astronomer of the time. Brahe's observations of Mars moving through the sky formed the basis for Kepler's work. Over the next several decades, Kepler proposed three laws that he believed governed the motions of all planets and moons.

Kepler's first law states that planets move around the sun in elliptical orbits, with the center of the sun located at one focus of the ellipse.

The second law states that an imaginary line drawn from the sun to a planet sweeps out equal areas in equal times (See Figure 1 below.). This means that planets move faster when they are closer to the sun.

The third law states that the time a planet takes to complete an orbit—its orbit period—is proportional to the radius of the orbit. More specifically, the period of rotation squared is proportional to the orbit's radius cubed, or $p^2 = k(r^3)$.

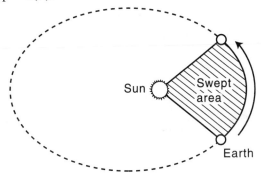

Figure 1. Area swept by an imaginary line

Before You Begin

Read the following guidelines for completing this test.

- Keep in mind that your teacher will be observing and grading your in-class behavior as well as your written responses. In particular, your teacher will be noting your ability to follow the given procedures, your adherence to classroom or laboratory safety, and your methods and reasoning in solving the problems.
- Try not to let what others are doing influence your work. Remember that a problem often has several acceptable solutions.
- Do not talk to other students unless you are working in a group. Talk only to members of your group and try not to disturb other students.
- Use only the materials provided.

Safety Alert

Scissors are sharp objects and should be handled carefully. Make sure that your group completes this activity at a safe distance from the other groups. Wear safety goggles.

M O D E R N   E A R T H   S C I E N C E

Unit 8: Performance-Based Test

Procedure

Step 1: Tie a loop in one end of the string and thread it through the spool, as shown in Figure 2. Tie several washers to the other end. Hook the spring scale onto the loop, and adjust the string until the washers are 1 m from the spool. Mark the string at this point.

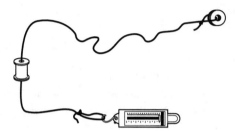

Figure 2. Experimental setup

Step 2: Hold the spool and the scale, and swing the washers in a circle just fast enough to keep the string taut (Figure 3). Practice swinging the washers until you can maintain a constant force on the scale. Record the amount of force in the data table on the next page. Continue to swing the washers while your group members do Step 3.

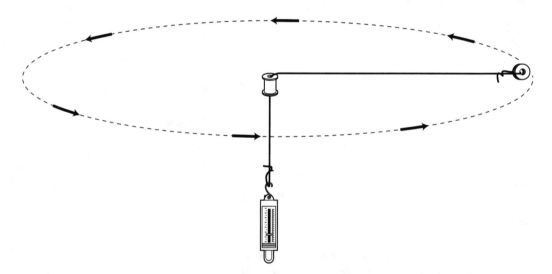

Figure 3. Swinging the weights

Step 3: Using a stopwatch, measure and record the time it takes for the washers to complete ten complete orbits. Record your results in the data table. Also calculate and record the period, or the time it takes to complete one orbit.

Step 4: Shorten the string by 15 cm and repeat Steps 2 and 3. Make sure that the force on the scale is the same as before. Record your data in the table. Repeat this step three more times, shortening the string by 15 cm each time.

M O D E R N   E A R T H   S C I E N C E

Unit 8: Performance-Based Test

Step 5: Let out the string to its original length. While swinging the washers, pull down and let up on the scale to create an elliptical orbit. Carefully observe the force on the string and the relative speed of the washers. Record your observations in the space provided below.

_____

_____

_____

_____

_____

Data Table

A	B	C	D	E
String length (radius), or r (cm)	Time/10 orbits (sec.)	Orbit period, or p (sec.)	p^2/r^3	Force (N)

Questions

Answer the following questions in the space provided. Support your answers by explaining your reasoning.

1. What do the washers, the spool, and the string represent in this experiment?

_____

_____

_____

_____

2. Use the space provided on page 59 to plot a graph of the data in columns C and D.
3. Kepler's third law states that the orbit period of a moon or planet is related to the average radius of its orbit. Do your results support this idea?

_____

_____

_____

_____

_____

M O D E R N   E A R T H   S C I E N C E

Unit 8: Performance-Based Test

4. Figure 4 (below) illustrates the elliptical orbit that you simulated in Step 5. On the diagram, mark the point at which the washers were moving the fastest with an F. Mark the point at which the washers were moving the slowest with an S.

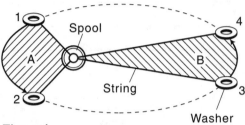

Figure 4.

5. In Figure 4, the areas swept by the string are equal (area of *A* = area of *B*). Based on this fact and your observations from Step 5, what can you infer about the time it takes the washers to go from *1* to *2* and from *3* to *4?* Explain your answer using Kepler's laws.

_____

_____

_____

_____

_____

_____

_____

6. Chasing and docking with another spacecraft, such as the shuttle docking with Russia's Mir space station, is very tricky. Using Kepler's laws, explain why such maneuvers are difficult.

_____

_____

_____

_____

_____

_____

_____

M O D E R N   E A R T H   S C I E N C E

Unit 8: Performance-Based Test

Answer Key to Performance-Based Tests

Contents

Materials List

Materials	[WARD'S Catalog No.]	Unit	Quantity (per student or group)
art gum, or soap eraser		6	1
beaker, small	[17 M 4020]	4	1
calculator	[27 M 3055]	1	1
clear plastic bottle with cap, 2–3 L		7	1
clear plastic container and lid	[18 M 1635] [17 M 2153]	3	1
clock or watch with a second hand	[15 M 1547]	4	1
colored pencils	[15 M 4690]	5	1 set
cork stoppers	[15 M 8366]	2	3
English-metric conversion table		1	1
fish tank, or other large transparent container	[21 M 5241]	2	1
fishing sinker	[350 M 0237]	2	3
grease pencil	[15 M 1155]	2, 4	1
matches	[15 M 9427]	7	1 box
mixture of silt, iron filings, and light-colored sand	[45 M 1982] [37 M 2312] [45 M 1983]	3	1/2 L
paper towel	[15 M 9844]	2	1 roll
pipe cleaner, or twist tie	[20 M 2109]	1	1
plastic fork		4	1
plastic window screening		4	1 m²
protractor (or magnetic compass)	[15 M 4067]	1	1
pushpin	[15 M 0507]	2	3
ruler	[15 M 4650]	1, 4, 5, 8	1
scissors	[14 M 0425]	2, 8	1
soft-lead pencil (No. 2 or softer)	[15 M 9816]	6	1
spring scale	[15 M 3741]	8	1
stopwatch	[15 M 0512]	8	1
stream table	[36 M 4211]	4	1
string	[15 M 9863]	2, 8	2–3 m
strip of thin balsa wood or polystyrene		2	1
tape	[15 M 1959]	2	1 roll
thread spool		8	1
tongue depressor, or similar wooden splint	[14 M 0103]	4	1 doz.
washer, or other small weight	[15 M 0030]	8	2–3
watering can, or spray bottle	[20 M 3700]	4	1

M O D E R N   E A R T H   S C I E N C E

Unit 1: Studying the Earth
Performance-Based Test: Using Topographic Maps

Objectives
Demonstrate your ability to read topographic maps accurately. Use the maps provided to solve problems in a variety of hypothetical situations.

Background
Maps are designed to convey information. All map elements, including the legend and title, contain useful information. However, reading or interpreting a map accurately is a skill that takes practice. It requires being able to identify a map's symbols and being able to determine direction and distance.

Maps are designed for a variety of purposes. The choice of map elements, such as the scale and contour interval, depends largely on the purpose of the map. Comparing maps whose scales, contour intervals, and systems of measurement differ is challenging but often necessary. Because map designs can vary widely, knowing what type of map is best suited for a particular problem is also necessary.

Before You Begin
Read the following guidelines for completing this test.
- Keep in mind that your teacher will be observing and grading your in-class behavior as well as your written responses. In particular, your teacher will be noting your ability to follow the given procedures, your adherence to classroom or laboratory safety, and your methods and reasoning in solving the problems.
- Try not to let what others are doing influence your work. Remember that a problem often has several acceptable solutions.
- Do not talk to other students unless you are working in a group. Talk only to members of your group and try not to disturb other students.
- Use only the materials provided.

Procedure
Use the assembled materials and the topographic maps on pages 8–10 to answer all the questions as completely as you can. The maps provided are portions of USGS maps that have been modified for your use.

M O D E R N   E A R T H   S C I E N C E

Unit 1: Performance-Based Test

Questions

Answer the following questions in the space provided. Support your answers by explaining your reasoning and showing calculations, if any.

1. Which of the three maps represents the largest area?

The Durango map. Because all three maps are roughly the same size, the map

drawn to the smallest scale represents the largest area.

2. Compare the areas of the cities of Comanche, Carlsbad, and Del Norte. Which city has the largest area?

Comanche. You may wish to check student calculations to distinguish incorrect

responses caused by arithmetic mistakes from those caused by a lack of

understanding of map scales.

3. Find the highest point on each map, and mark these points with an *A*. Which map contains the highest point?

See the maps on pages 72–74 for the location of the highest point on each map.

The Durango map contains the highest point.

4. If you were using a compass in Comanche, Texas, in what direction would the compass needle point with respect to true north?

The compass needle would point to magnetic north, or 7.5° east of true north.

5. Which map is the least distorted? Why?

The Carlsbad map is the least distorted. Distortion results from trying to

represent the earth's curved surface on a two-dimensional map. A map covering

a small area is less distorted because the surface it maps is closer to being flat.

M O D E R N   E A R T H   S C I E N C E

Unit 1: Performance-Based Test

6. Briefly describe the view you might see if you were looking east from Del Norte?

   <u>Answers may vary. A typical response would describe a relatively flat area with a</u>
   <u>railroad and the Rio Grande in the foreground. Students may also infer from the</u>
   <u>presence of canals and irrigation ditches that this area is farmland.</u>

   looking west?

   <u>Answers may vary. A typical response would describe the Rio Grande Valley</u>
   <u>running from west to east, with mountains on either side of the valley.</u>

   looking north?

   <u>Answers may vary. A typical response would describe the Rio Grande in the</u>
   <u>foreground, with mountains to the west and flat farmland to the east.</u>

7. Locate the stream nearest the radio tower west of Carlsbad, at the southeastern edge of the Ocotillo Hills. In which direction does the stream flow? Explain your answer.

   <u>The stream flows roughly from north to south. The stream's direction is indicated</u>
   <u>by the V-shaped contour lines that point upstream and by the overall decrease</u>
   <u>in elevation from north to south.</u>

8. Fill in the missing information in the table below. (The contour intervals on both the Carlsbad map and the Durango map are shown in feet.)

Map name	Verbal scale	Map scale	Contour interval
Carlsbad	1 cm = 240 m	1:24,000	20 ft.
Comanche	1 cm = 1 km	1:100,000	10 m
Durango	1 cm = 2.5 km	1:250,000	200 ft.

M O D E R N   E A R T H   S C I E N C E

Unit 1: Performance-Based Test

9. Maps with different scales are useful for different purposes. Choose which map would be the most useful for each of the following activities. Explain your choices.
   a. planning a five-day hiking trip
   b. deciding where to build a new industrial plant
   c. picking a site for a new housing development

The Durango map would be best for planning a hiking trip because it covers the

largest area and offers the most choices of five-day hikes. The Carlsbad map

would be best for deciding where to build an industrial plant because it shows in

detail the roads, waterways, and unoccupied land that a plant might need. The

Comanche map would be best for picking a housing site because it shows all of

the areas closely surrounding the major population center while still providing

some details.

10. Can you determine distances with a map based on the metric system if your ruler is based on the English system of measurement? Explain.

Yes. Map scales can always be expressed as a ratio of map distance to actual

distance. This ratio remains the same regardless of which system of

measurement is used.

11. On the appropriate maps, locate the roads that connect the following places:
   • the western edge of Carlsbad and the buildings in the state park
   • the eastern edge of Comanche and the town of Blanket
   • Del Norte and Monte Vista
   If it costs $600,000 to widen 1 km of road, which road would cost the most to widen? Approximately how much would it cost? (Hint: Use a pipe cleaner or twist tie to measure the curved parts of the roads.)

The Del Norte road is the longest and therefore would cost the most to widen. It

would cost about $9.6 million (16 km × $600,000/km = $9,600,000).

Name _____   Class _____   Date _____

Unit 1: Performance-Based Test

12. Which of the roads listed in Question 11 is the steepest?

The Carlsbad road is the steepest. It rises about 250 ft. in about 2.6 km. The

Comanche road rises about 60 m (or about 200 ft.) in about 14 km. The Del Norte

road is essentially flat.

_____

_____

_____

_____

13. Find the radio tower just west of Carlsbad. Is this the best location for the radio tower? If not, select a new location for the tower and mark it with a point labeled *B*. Explain why your new location, if any, is better.

Answers may vary. See the map on page 72 for a sample location. Students

should recognize that the best place for the tower is on the highest point of land.

Putting the tower on the highest point would increase the tower's range. Some

students may argue that relocating the tower on higher ground would require

raising the height of the tower to service western Carlsbad because radio

signals travel along line-of-sight vectors.

_____

_____

_____

14. A pilot radios you and tells you that she has crashed her plane after traveling 52 km from Monte Vista, Colorado. She reports being on a heading of 270° according to her magnetic compass. Magnetic declination for the Durango map is 12° E. Place a point labeled *B* where you would expect to find the pilot.
(See the map on page 73 for the solution.)

RW material copyrighted under notice appearing earlier in this work.

71

Unit 1: Performance-Based Test

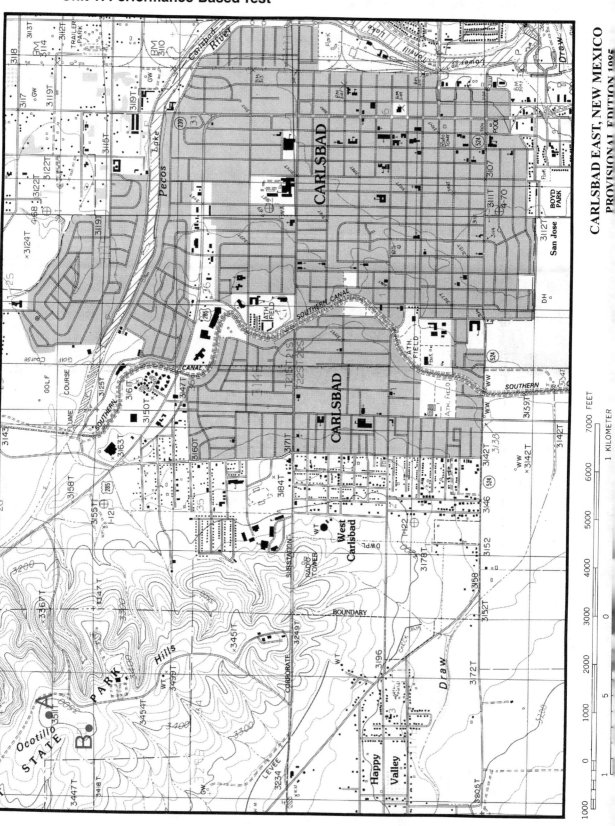

CARLSBAD EAST, NEW MEXICO
PROVISIONAL EDITION 1985

Unit 1: Performance-Based Test

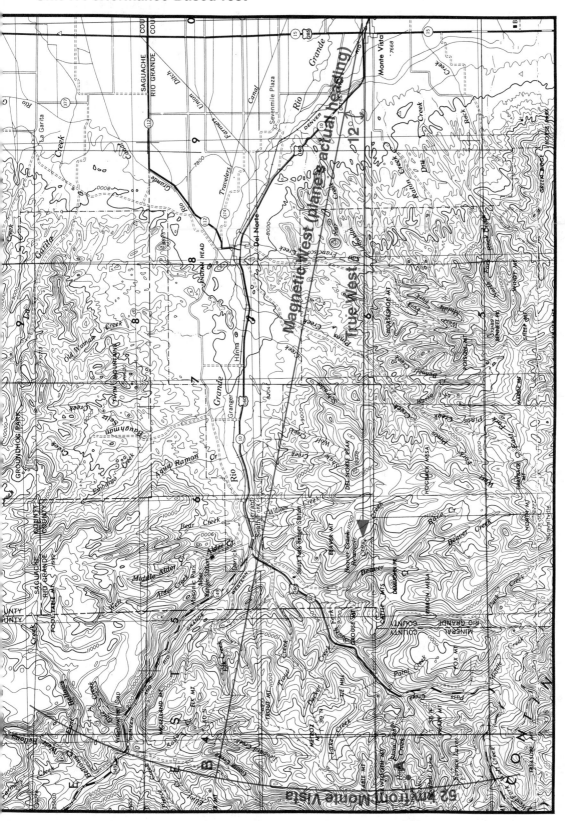

DURANGO, COLORADO
1945

Scale 1:250,000

M O D E R N   E A R T H   S C I E N C E

Unit 1: Performance-Based Test

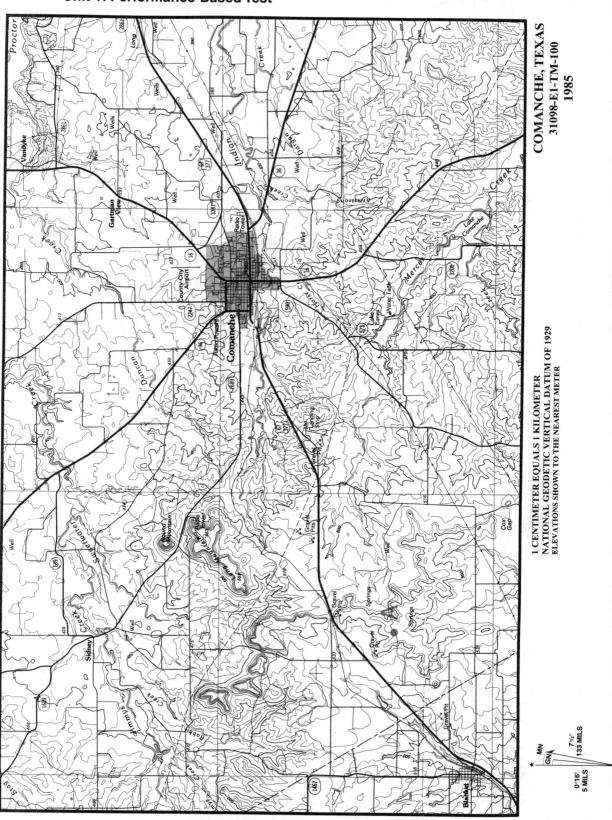

COMANCHE, TEXAS
31098-E1-TM-100
1985

1 CENTIMETER EQUALS 1 KILOMETER
NATIONAL GEODETIC VERTICAL DATUM OF 1929
ELEVATIONS SHOWN TO THE NEAREST METER

M O D E R N   E A R T H   S C I E N C E

Unit 2: The Dynamic Earth
Performance-Based Test: Making Waves

Objectives
Use a water tank to try to create the three main types of seismic waves. Modify a rigid surface structure to make it less susceptible to earthquake damage.

Background
The energy released by earthquakes moves through the earth in the form of mechanical waves. Earthquakes generate three types of waves: compression, or P, waves; shear, or S, waves; and surface, or L, waves. Unlike electromagnetic waves, mechanical waves require a medium through which to travel. Differences in the medium greatly affect mechanical waves. By studying seismic waves, scientists can learn a great deal about the composition of the earth.

Before You Begin
Read the following guidelines for completing this test.
- Keep in mind that your teacher will be observing and grading your in-class behavior as well as your written responses. In particular, your teacher will be noting your ability to follow the given procedures, your adherence to classroom or laboratory safety, and your methods and reasoning in solving the problems.
- Try not to let what others are doing influence your work. Remember that a problem often has several acceptable solutions.
- Do not talk to other students unless you are working in a group. Talk only to members of your group and try not to disturb other students.
- Use only the materials provided.

Safety Alert
Any water spilled on the floor should be mopped up immediately to prevent slipping. Pushpins and scissors are sharp objects and should be handled carefully. Tanks should also be handled carefully so as not to break them.

Procedure
Step 1: Set up the tank, sinkers, and cork stoppers as shown in Figure 1 below. The first stopper should float on the surface and should have about 2 cm of extra string. The second stopper should be about two-thirds of the way from the bottom, and the third stopper should be about one-third of the way from the bottom. Be sure to draw the line on the side of the tank as shown to help you observe the motion of the corks.

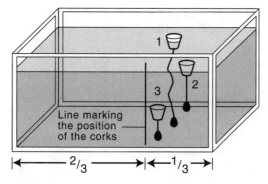

Figure 1

Unit 2: Performance-Based Test

Step 2: Make a loose fist near the bottom of the tank and point your thumb toward the stoppers (Figure 2). Quickly and firmly clench your fist without moving your arm. Record any motion of the three stoppers. You may have to place your hand closer to the stoppers to achieve any noticeable results.

Students should notice motion in the submerged stoppers but not in the floating

stopper. The submerged stoppers should move back and forth.

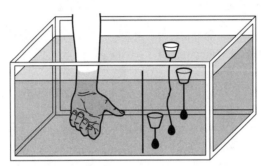

Figure 2

Step 3: Repeat Step 2, but with your thumb pointed toward the bottom of the tank. Record your observations.

Students should notice motion in the surface stopper but not the submerged

stoppers. The surface stopper should move up and down as well as back and forth.

Step 4: Open your fist and place your hand halfway into the water with your palm facing the stoppers (Figure 3). Move your hand from side to side. Record your observations.

Students should notice little or no motion in any of the stoppers.

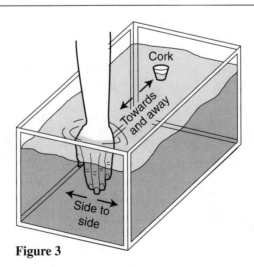

Figure 3

M O D E R N   E A R T H   S C I E N C E

Unit 2: Performance-Based Test

Step 5: Repeat Step 4, but this time move your hand toward and away from the corks. You should generate several waves at a time. Record your observations.

Students should notice motion in the surface stopper but not the submerged

stoppers. The surface stopper should move up and down as well as back and

forth.

Step 6: Empty the tank halfway and float the strip of balsa wood or polystyrene on the water (Figure 4). Make waves as in Step 5. Make the waves as large and as rapidly as you can without spilling the water. Observe the strip from the side. Record your observations, noting especially the motion at the ends and the middle of the strip.

Students should notice that the strip is longer than an individual wave. If the

waves are tall enough, students should notice that the ends and the middle of

the strip are clear of the water at different times.

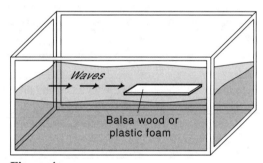

Figure 4

Questions

Answer the following questions in the space provided. Support your answers by explaining your reasoning.

1. Which type of wave did you create in Step 2? Explain.

   Compression waves. Opening and closing a fist created the compressions that

   caused the submerged stoppers to move back and forth.

2. Which type of wave did you create in Step 3? Explain.

   Surface waves. Only the surface stopper moved. Students should realize that

   they created a compression wave by clenching their fist but that the

   compression wave changed to a surface wave.

M O D E R N   E A R T H   S C I E N C E

Unit 2: Performance-Based Test

3. Which type of wave did you try to create in Step 4? Explain your observations.

   Step 4 was an attempt to create shear waves. None of the stoppers moved
   because shear waves move only through solids.

4. Which type of wave did you create in Step 5? Explain.

   Surface waves. Only the surface stopper moved.

5. Based on the motion of the surface stopper, how do you think the particles of the medium move to create surface waves?

   The particles of the medium move in circles, as shown by the up-and-down
   motion of the stopper combined with its back-and-forth motion.

6. If a seismograph doesn't record any shear waves from an earthquake, what can you infer about the medium through which the seismic waves passed?

   The earthquake waves encountered a liquid before reaching the recording
   station.

7. From your observations of the floating strip in Step 6, explain why surface waves do so much damage to buildings that cover large areas.

   If a building covers a large area, the building is affected by multiple waves
   simultaneously. This causes different parts of the building to move in different,
   often opposite, directions, and that can tear the building apart.

8. Suppose that the floating strip is the foundation of a building. Use the assembled materials to make the strip better able to withstand an earthquake. Briefly explain your modifications.

   Solutions may vary. Accept all reasonable solutions that result from thorough
   analysis. Students may choose either to reinforce the strip so it doesn't bend or
   break or to cut the strip into smaller sections joined together by string.

Unit 3: Composition of the Earth

Performance-Based Test:
Prospecting for Ores and Oil

Objectives

Use your knowledge of the physical and chemical properties of ores to select possible mining sites. Analyze graphs of oil-well production to identify profitable wells.

Background

Today's societies are increasingly dependent on sources of ore minerals. The search for earth's minerals is no longer the simple one- or two-person operation of early prospectors. It is highly organized and involves the latest scientific methods and equipment. However, as with early prospecting, a knowledge of the properties of ore minerals is required for success.

Exploring for fossil fuels has also advanced as demand has increased. New methods have been developed not only to find these deposits but also to gauge potential production. One method of monitoring an oil well's productivity is to plot the history of the well's oil production. These histories indicate when pumping a well is no longer economically viable and how much oil remains in the well.

Before You Begin

Read the following guidelines for completing this test.

- Keep in mind that your teacher will be observing and grading your in-class behavior as well as your written responses. In particular, your teacher will be noting your ability to follow the given procedures, your adherence to classroom or laboratory safety, and your methods and reasoning in solving the problems.
- Try not to let what others are doing influence your work. Remember that a problem often has several acceptable solutions.
- Do not talk to other students unless you are working in a group. Talk only to members of your group and try not to disturb other students.
- Use only the materials provided.

Procedure

Fill the clear container about one-quarter full with a mixture of sand, silt, and particles of heavy metal, such as iron filings or crushed galena. Add water until the container is almost full. Swirl the contents of the container to thoroughly mix the solids and water. Observe the mixture as you continue to agitate the container. Gradually slow down until you are causing the liquid to barely move.

M O D E R N   E A R T H   S C I E N C E

Unit 3: Performance-Based Test

Questions
Answer the following questions in the space provided. Support your answers by explaining your reasoning.

Section I: The Location of Mineral Ores

1. What did you observe about the order in which the different materials settled to the bottom of the container?

   The denser materials settled first. The heavy metal settled first, followed by the

   sand and then the silt.

   _____

   _____

2. How do you think the speed of the water as you swirled and agitated the jar affected the order in which the different materials settled?

   The faster the water was moving, the greater the quantity and density of the

   sediment in suspension.

   _____

   _____

3. How might the speed of the water affect heavy minerals, such as gold, silver, or tin, as they are washed down a stream or onto a beach?

   Heavy minerals, such as gold and silver, can be carried by fast-moving streams

   but are dropped when the water slows down. This leads to placer deposits where

   the stream slows down abruptly, such as at the base of waterfalls and rapids, the

   inside of stream bends, and even the wave-washed zone of some beaches.

   _____

   _____

4. On the map on page 24, which letters represent locations where placer gold deposits might be found? Why do you think there may be gold there? (Hint: There must be a source of gold in order for a placer deposit to form.)

   Answers may vary. Accept all reasonable responses that are well supported. The

   igneous intrusion is the most probable source of the area's gold. Placer deposits

   are likely at *N* (Hope's Rapids), *L* (Frostbite Falls), *F* (Mule Shoe Slough),

   D (Burnt Boat Delta), and areas *I, G,* and *H*. These are places where the water

   would slow down and drop gold.

   _____

   _____

M O D E R N   E A R T H   S C I E N C E

Unit 3: Performance-Based Test

5. Which letters on the map represent locations where you might find minerals formed through contact metamorphism? Explain.

Answers may vary. Accept all reasonable responses that are well supported.

Typical responses might include L, U, T, W, and X. These are areas of

metamorphic rock that are in contact with the igneous intrusion. Areas L, W, and

X are speculative because they are covered by soils.

_____

_____

6. While still in its original rock, gold is often associated with sulfide minerals. Using the reported findings on the map, pick the best locations to prospect for this type of deposit. Explain your choices.

Answers may vary. Accept all reasonable responses that are well supported. A

typical response would be area M and the adjacent limb of the intrusion. If the

presence of sulfide is used to locate gold, areas with the highest concentration

of sulfur would be best. Dangerfield Draw and Youngman Creek both drain areas

high in sulfur minerals.

_____

_____

_____

Section II: The Production Rates of Oil Wells

The graphs on pages 25–26 plot the production histories of six wells. Refer to these graphs to answer the remaining questions. Remember to justify all of your answers.

7. Which well has produced the most oil to date?

Holt #3 has a total production of 3.5 billion cubic feet of oil, making it the

fastest-producing of all the wells.

_____

_____

8. Which well is producing at the highest rate?

Holt #6 is currently producing at the highest rate. Give full credit to students

who choose Holt #3 if their reasoning is that the production of Holt #6 is not

continuous and is highly undependable.

_____

_____

M O D E R N   E A R T H   S C I E N C E

Unit 3: Performance-Based Test

9. Which well is producing at the lowest rate?

Holt #1 has the lowest production rate. It is producing just over 1 million cubic feet of oil per day.

10. Which well will probably be the first to be abandoned as unprofitable?

Holt #5. Although its production rate is not closest to the abandonment rate, the steep slope of the curve suggests that this well has less than a year left. Holt #1, although producing closer to the abandonment rate, has more time left, as shown by its flatter rate curve.

11. Which well probably has the most oil left to produce?

Holt #4. Although the well's production rate is not the highest, the rate curve is almost flat. This indicates that the well has a long time left before it will be abandoned. In its lifetime, it will probably outproduce the other wells.

12. Suppose that you are an oil company's expert in buying existing oil wells. The six wells whose production rates are graphed on pages 25–26 are up for sale. List your buying preferences in order from best to worst. For each well, explain its ranking.

Well name: Holt #3

Holt #3 shows a slightly steeper drop in its production curve than Holt #4. This means that Holt #3 will not produce as long or as much as Holt #4. However, its higher rate of production will give a quicker return on your investment than Holt #4.

Well name: Holt #4

This well, although producing at a moderate rate, may continue to produce oil for a very long time. The small slope of its production curve suggests that Holt #4 is far from abandonment and will produce a large amount of oil over its lifetime.

M O D E R N   E A R T H   S C I E N C E

Unit 3: Performance-Based Test

Well name: Holt #2

A faster drop in production and a lower rate of production suggest that this well

will produce less oil than the first two wells.

_____

_____

_____

_____

Well name: Holt #6

Although its production rate is sporadic, total production is high, and the well

appears to have a long life ahead of it. The total production curve is increasing,

which shows that periods of inactivity do not seem to be hindering the well's

overall production. More information is needed to understand why the well

periodically fails to produce.

_____

Well name: Holt #1

A low production rate that is very close to the lowest economic rate makes this

well unattractive. Neither Holt #1 nor Holt #5 is very productive.

_____

_____

_____

_____

Well name: Holt #5

A low production rate that is very close to the lowest economic rate makes this

well unattractive. Neither Holt #1 nor Holt #5 is very productive.

_____

_____

_____

_____

M O D E R N   E A R T H   S C I E N C E

Unit 4: Reshaping the Crust

Performance-Based Test: Controlling Erosion

Objectives

Test several methods of soil conservation and evaluate their effectiveness. Use your results to solve problems in soil conservation.

Background

One of our most important natural resources is high-quality topsoil. Without this resource, we could not grow food. As valuable as this resource is, it gets very little public attention. While it is true that farmers and ranchers share the major responsibility for protecting this resource, we all suffer from the results of soil erosion.

Millions of dollars are spent annually dredging eroded soil from rivers and bays. This eroded soil not only threatens our use of waterways but also decreases the available wetlands and water sources used by wildlife. Even the soil eroding from our lawns may cause problems by clogging storm drains and causing flooding.

Before You Begin

Read the following guidelines for completing this test.
- Keep in mind that your teacher will be observing and grading your in-class behavior as well as your written responses. In particular, your teacher will be noting your ability to follow the given procedures, your adherence to classroom or laboratory safety, and your methods and reasoning in solving the problems.
- Try not to let what others are doing influence your work. Remember that a problem often has several acceptable solutions.
- Do not talk to other students unless you are working in a group. Talk only to members of your group and try not to disturb other students.
- Use only the materials provided.

Safety Alert

Water spills must be mopped up immediately to prevent slipping.

Procedure

Step 1: Create a smooth slope of moist sand in a stream table. With a grease pencil, trace the angle of the slope on the side of the stream table. Water the upper part of the slope evenly until a stream forms. Record the height from which you pour the water, the amount of water you use, and the time it takes to form a stream. You may need several containers of water to obtain the desired effect.

Step 2: Use a small beaker to measure the amount of sand that eroded. Record this amount in the table on the next page.

Step 3: Restore the slope to its original shape using the line you marked with the grease pencil as a reference. Use a fork to plow furrows perpendicular to the angle of the slope. Create a deeper furrow every so often, as shown in Figure 1 on the next page. Add water to the slope exactly as you did in Step 1, and record the amount of eroded sand in the data table.

Unit 4: Performance-Based Test

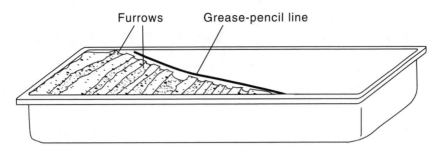

Furrows   Grease-pencil line

Figure 1

Step 4: Restore the slope. Place a plastic screen on the slope. Water the slope, and record the amount of eroded sand.

Step 5: Using the diagram below as a guide, use tongue depressors to create a series of terraces on the slope. Make sure you align the tops of the tongue depressors with the line you marked with the grease pencil. Also be sure that the tongue depressors are slightly higher than the sand behind them. Water the slope, and record the amount of eroded sand.

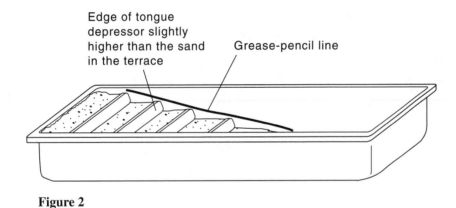

Edge of tongue depressor slightly higher than the sand in the terrace   Grease-pencil line

Figure 2

Step 6: If time allows, repeat some or all of the setups at steeper angles.

Data Table

Conservation method	Amount of eroded sand

The unprotected slope will lose the most sand. The terraced slope will lose the least sand. The effectiveness of the screen and countour plowing may be nearly identical and may depend on the angle of the slope the student selects.

M O D E R N   E A R T H   S C I E N C E

Unit 4: Performance-Based Test

Questions

Answer the following questions in the space provided. Support your answers by explaining your reasoning.

1. What method of soil conservation does the screen represent?

The screen represents the conservation of soil through the use of plant cover.

_____

2. Why was it important to water the slope the same way for each method of soil conservation?

Watering the slope the same way each time kept the force of the water constant.

This ensured that any differences in the amount of eroded sand resulted from

differences in the method of soil conservation.

_____

3. According to your results, which method of soil conservation protected the soil the best? Why do you think this method was the most effective?

Results may vary, depending on the angle of the slope and the force of the water.

In general, the terraced slope should perform best and the unprotected slope

should perform worst. The greater the force of the water flow, the greater the

amount of soil the water can carry and the more effective its sediment load is at

further eroding the slope. The terraced slope is the most effective at reducing

the force of the water flow.

_____

4. How would you expect your results to be different if the angle of the slope were steeper? Explain.

A steeper slope would result in an overall increase in the amount of soil erosion.

Some conservation methods, such as contour plowing, might be ineffective if

the slope is too steep.

_____

5. Which conservation method would you recommend for gently sloping hills?

Contour plowing would probably be adequate for gentle slopes because the rate

of water flow would be low.

_____

_____

M O D E R N   E A R T H   S C I E N C E

Unit 4: Performance-Based Test

6. Which conservation method would you recommend for steeply sloping hills?

Terracing would probably be the best choice for steep slopes because it is the most effective method for slowing down the water flow. Also, terracing may be the only way to make farming possible if the slope is extremely steep.

7. Try to think of what objections a farmer might have to each of the following methods of soil conservation.

contour plowing

Contour plowing may be difficult, or even dangerous, because it often requires plowing perpendicular to the angle of the slope to prevent gullying.

plant cover

Relying on plant cover requires periodically leaving fields fallow or carefully selecting different types of plants that can be used in crop rotation or strip-cropping.

terracing

Terracing is labor-intensive and expensive because it requires building walls and leveling the terraces. It is also difficult to operate equipment in terraces, and the terraces may retain too much water for some crops.

8. Which method(s) would also work against wind erosion? Explain your choice(s).

Wind erosion occurs primarily through saltation, a process by which the particles of soil skip along the ground. If rows are plowed at right angles to the prevailing wind, contour plowing should be able to protect against wind erosion. Plant cover would also reduce wind erosion by trapping windblown soil particles.

M O D E R N   E A R T H   S C I E N C E

Unit 5: The History of the Earth

Performance-Based Test: Writing Geologic History

Objectives
Use geologic columns from several areas to construct regional geologic columns. Make inferences about the region's geologic history by comparing the regional and local columns.

Background
Much of the history of how the earth's landscapes and environments have changed lies in the rock record. However, no single area on the earth contains a record of all geologic time. To compile a complete history, geologists use clues such as fossils, rock type, and mineral content to combine local geologic columns into larger regional ones.

   Once a regional geologic column is assembled, geologists try to decipher the meaning of gaps or changes in rock type across the region. They also try to reconstruct the landscape and environment of the region at different times. In doing so, geologists gain an understanding of the processes responsible for the earth's complex geologic history.

Before You Begin
Read the following guidelines for completing this test.

- Keep in mind that your teacher will be observing and grading your in-class behavior as well as your written responses. In particular, your teacher will be noting your ability to follow the given procedures, your adherence to classroom or laboratory safety, and your methods and reasoning in solving the problems.
- Try not to let what others are doing influence your work. Remember that a problem often has several acceptable solutions.
- Do not talk to other students unless you are working in a group. Talk only to members of your group and try not to disturb other students.
- Use only the materials provided.

Procedure
Step 1: In Figure 1 below, note how a geologist might use a geologic column to record the different sediments present in a shoreline.

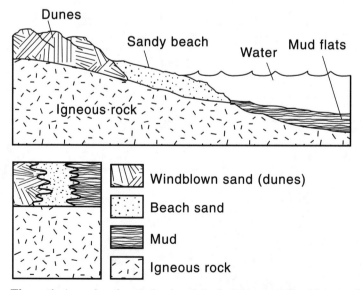

Figure 1. A modern beach (top) and the beach represented in a geologic column (bottom)

Unit 5: Performance-Based Test

Step 2: Use the geologic columns on page 39 to construct two regional geologic columns in the blank columns below. Notice that the geologic columns on page 39 are divided into a western and an eastern group. Create one regional geologic column for each group of three, local, geologic columns. The thickness of a rock layer in a regional column should equal the maximum thickness attained by that rock layer in any of the three corresponding columns.

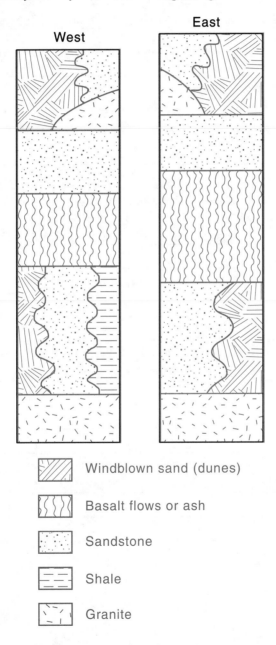

West   East

Windblown sand (dunes)

Basalt flows or ash

Sandstone

Shale

Granite

Step 3: The geologic columns on page 39 correspond to the locations labeled on the base map on page 40. On the base map, briefly sketch the surface features of the landscape suggested by the oldest sedimentary layer. Be sure to label each feature and include shorelines to distinguish aquatic environments from terrestrial environments. (See the sample completed map on page 92.)

M O D E R N   E A R T H   S C I E N C E

Unit 5: Performance-Based Test

Questions

Answer the following questions in the space provided. Support your answers by explaining your reasoning.

1. How do you explain the distribution of the geologic columns on the base map?

   Geologists gather stratigraphic data wherever it is available. The locations of the

   geologic columns probably represent natural exposures of rock that are

   randomly spaced.

2. What does the wavy surface of the lowest rock layer in this region represent?

   The wavy surface represents an unconformity where rock either was not

   deposited or was removed by erosion.

3. What was the source of the sediments that make up the lowest sedimentary layer in the geologic columns? Justify your answer.

   Answers may vary. Accept all reasonable responses. The sediments in this

   rock layer may have come from the erosion of terrestrial sediments, the

   remains of aquatic organisms, or both. Students may also suggest that the

   sediments resulted from glacial erosion.

4. Suggest an index fossil for the geologic columns of this region.

   The fossil snail in the upper sandstone layer appears in only that one layer and

   only for a short time. If this were true over a large area, then the snail would be a

   good index fossil.

5. Explain the existence of a basalt layer in all of the geologic columns.

   The basalt layer is an extrusive igneous formation, which indicates volcanic

   activity or rifting.

M O D E R N   E A R T H   S C I E N C E

Unit 5: Performance-Based Test

6. What type of geological feature is evident in geologic column *C?* What is its relative age?

Column *C* shows a normal fault that occurred after the basalt layer formed but

before the sandstone layer formed.

_____

_____

7. What type of igneous rock formation is evident in the upper part of geologic columns *C* and *D?* What is its relative age?

The top igneous layer in columns *C* and *D* is a granitic intrusion (sill). It is

probably younger than the rock above it because fossil snails appear in the

sandstone above and below the intrusion. It is also younger than the rock layers

below because it cuts through all of these rock layers.

_____

8. What type of tectonic-plate interaction is suggested by the top layer in column *E?*

The trench sediments suggest that oceanic crust is being subducted by

continental crust to the east of column *E.*

_____

9. Summarize the geologic history of the region, and cite specific evidence in the geologic record to support your interpretation.

Answers may vary. Accept all reasonable responses that are well supported.

Typical answers might describe the following sequence of events: erosion of the

bottom layer of granite, indicated by the unconformity; formation of a shallow

sea in the center of the region, indicated by the sandstone and shale layers in

columns *B–E,* and surrounded by sand dunes on either side, indicated by

columns *A* and *F;* volcanism and possible rifting, indicated by the basalt layer in

all of the columns; and finally the subduction of oceanic crust, indicated by the

trench sediments in column *E.*

_____

_____

_____

_____

M O D E R N   E A R T H   S C I E N C E

Unit 5: Performance-Based Test

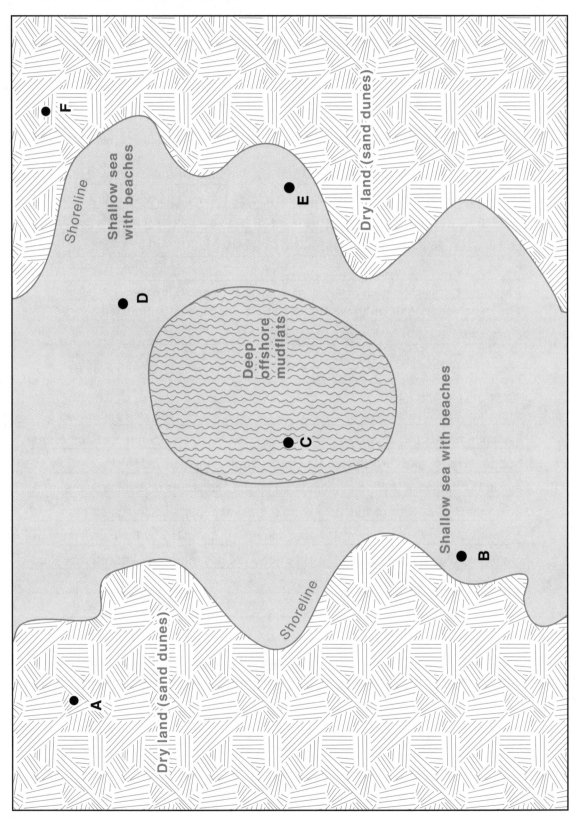

Base Map

F

Shoreline

Shallow sea
with beaches

Dry land (sand dunes)

E

D

Deep,
offshore
mudflats

C

Shallow sea with beaches

B

Shoreline

Dry land (sand dunes)

A

M O D E R N E A R T H S C I E N C E

Unit 6: Oceans

Performance-Based Test:
Mapping the Ocean Floor

Objectives

Use data on water depths to draw a topographic map of the ocean floor. Use your completed map to identify underwater features, and relate the features to ocean environments and plate tectonics.

Background

Humankind has long sought to map the sea floor. The first attempts focused on shallow areas and shorelines to prevent ships from running aground and sinking. These early maps were made with a common tool: a plumb bob, or a weight on a string.

Today, depth measurements are largely made with *sonar,* an acronym for *so*und *n*avigation *a*nd *r*anging. Sonar relies on the travel time of a sound signal to determine distance, or range. The technique involves transmitting a sound signal underwater and timing how long it takes the signal to return after being reflected by the ocean floor. After adjusting the speed of sound for factors such as water temperature and density, the time is converted to a depth. Almost all naval vessels and some civilian ships operate sonar equipment continuously and report their data to the Navy.

Before You Begin

Read the following guidelines for completing this test.

- Keep in mind that your teacher will be observing and grading your in-class behavior as well as your written responses. In particular, your teacher will be noting your ability to follow the given procedures, your adherence to classroom or laboratory safety, and your methods and reasoning in solving the problems.
- Try not to let what others are doing influence your work. Remember that a problem often has several acceptable solutions.
- Do not talk to other students unless you are working in a group. Talk only to members of your group and try not to disturb other students.
- Use only the materials provided.

Procedure

Examine the sonar data on the base map on page 46. Select an appropriate contour interval, and contour the map as quickly and as cleanly as you can. Answer Question 1 to convert some of the data from seconds to meters before you begin. Make your lines very faint at first because you will probably be erasing often. Do not darken any lines until you are sure that you are satisfied with their placement. Remember to label the contour lines.

M O D E R N   E A R T H   S C I E N C E

Unit 6: Performance-Based Test

Questions

Answer the following questions in the space provided. Support your answers by explaining your reasoning.

1. Use the graph in the upper right corner of the map on page 46 to determine the ocean depth at points *H, I,* and *J.*

   H = −6500 m; I = −6500 m; J = −5200 m

2. In the graph on page 46, why does the actual curve differ from the ideal curve?

   The travel time of a sonar signal is not a direct function of ocean depth because the speed of sound in ocean water is affected by factors that change with depth. These factors include temperature, density, and pressure.

3. Explain your choice of contour interval for the map.

   Contour interval may vary with the student. A contour interval of 1,000 m will produce a map with enough detail without being overly busy and difficult to read.

4. At which of the locations *A–G* would you expect to find the coldest water?

   Ocean temperature decreases down to a depth of about 1,200 m. The coldest water should be found at locations *A–E* because these locations are below 1,200 m. *F* and *G* are probably above the thermocline and should have warmer water.

5. At which of the locations *A–G* would you expect to find water with low salinity?

   Port City is a likely source of fresh water, so points *A, F,* and *G* are probably less salty.

6. What type of ocean-floor feature is located at each of the lettered locations *A–G?*

   A—trench; *B*—seamount or guyot; *C*—transform fault; *D*—abyssal plain; *E*—mid-ocean ridge; *F*—spit; *G*—continental shelf

Unit 6: Performance-Based Test

7. The data are not evenly spaced around the map. How do you explain the distribution of the data?
   Very few ships are dedicated to sonar measurements for research purposes. The
   majority of sonar readings are collected by ships on other missions. For
   practical reasons, water depth is most intensively sampled along shipping
   routes.

8. At which of the locations *A–G* would the fishing for bottom-dwelling organisms be best? Why?
   Locations *F* and *G* are probably the best because they both contain sublittoral
   zones. These zones are ideal for benthos because they have abundant sunlight,
   a fairly constant temperature, and relatively low pressure.

9. What does the feature at location *E* suggest about tectonic activity in the area?
   The mid-ocean ridge at location *E* suggests that there is a divergent plate
   boundary nearby where new oceanic crust is being created by upwelling magma.

10. What do you see on the map that would support a change in sea level over time? Explain your answer.
    Answers may vary and may include the straight shoreline showing past
    emergence; the drowned valley at the mouth of Port City Bay showing a rise in
    sea level; and the drowned Port City Bay indicating some submergence.

11. If you were to select an area on the base map for more-detailed depth readings, where would it be? Explain your choice.
    Answers will vary. Accept all reasonable responses that are well supported.
    Students may choose known features, such as the trench or the mid-ocean
    ridge, or they may choose areas where few depth readings have been taken so
    far.

MODERN EARTH SCIENCE

Unit 6: Performance-Based Test

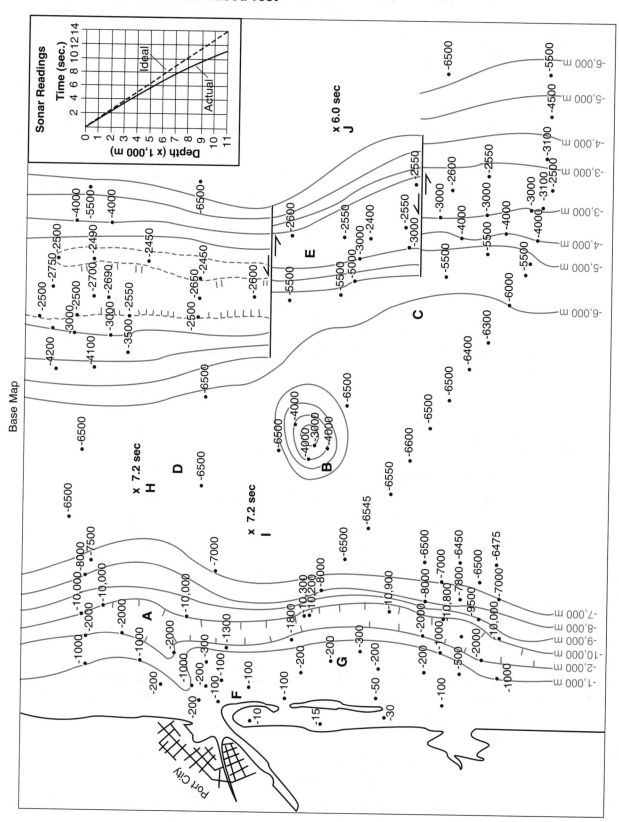

MODERN EARTH SCIENCE

Unit 7: Atmospheric Forces
Performance-Based Test: Creating Clouds

Objectives
Create clouds in a bottle to identify the conditions required for cloud formation. Compare your experimental conditions to conditions in the atmosphere.

Background
Clouds are a common sight. The rain they bring can range from a welcome relief to a minor inconvenience to a natural disaster. Because we are dependent on and often at the mercy of rain, humankind has long tried to control clouds and rain but without much success. A key to control is our understanding of the natural processes involved. Although our understanding of how clouds and rain form is not complete, we now know enough to make some crude attempts at controlling clouds and rain.

Before You Begin
Read the following guidelines for completing this test.
- Keep in mind that your teacher will be observing and grading your in-class behavior as well as your written responses. In particular, your teacher will be noting your ability to follow the given procedures, your adherence to classroom or laboratory safety, and your methods and reasoning in solving the problems.
- Try not to let what others are doing influence your work. Remember that a problem often has several acceptable solutions.
- Do not talk to other students unless you are working in a group. Talk only to members of your group and try not to disturb other students.
- Use only the materials provided.

Safety Alert
Matches are a fire hazard. Be sure you are familiar with the safety guidelines for working around open flames. Any water spilled on the floor should be mopped up immediately to prevent slipping. Loosely closed bottles may launch their caps when squeezed. Wear safety goggles.

Procedure
Step 1: Fill the bottle about half full with water at room temperature. Drop a lit match into the bottle, and screw the cap on tightly.

Step 2: Make sure the top of the bottle is not pointed at anyone, including yourself. Squeeze the sides of the bottle as hard as you can and quickly release it. Do this several times until a cloud forms. Record the number of squeezes in the table on the next page. Empty the bottle, and squeeze it several times to clear the smoke from the bottle.

Step 3: Repeat Steps 1 and 2 with warm water, and record your results.

Step 4: Repeat Steps 1 and 2 with cold water, and record your results.

Step 5: Repeat Steps 1–4 but do not use any matches. Record your results.

M O D E R N   E A R T H   S C I E N C E

Unit 7: Performance-Based Test

Data Table

Water temperature	Match (yes or no)	Number of squeezes
Room temperature	Data will vary, depending on water temperature, amount of water, and student strength. However, the warm water should require fewer squeezes than the cold water. The steps without the matches should require more squeezes than their counterparts with the match. In some cases, students may not be able to get a mist to form. Reassure students that this result may be valid and have them continue with their work.	
Warm		
Cold		
Room temperature		
Warm		
Cold		

Questions

Answer the following questions in the space provided. Support your answers by explaining your reasoning.

1. How did the water temperature affect the number of squeezes needed to create a cloud? Explain.

   The number of squeezes was inversely proportional to the water temperature.

   The warmer the water was, the fewer the number of squeezes were needed

   because the warmer water increased the amount of water vapor in the bottle.

2. How did using matches affect the number of squeezes needed to create a cloud? Explain.

   Using matches reduced the number of squeezes needed. Students should

   understand that the smoke from the matches provided condensation nuclei

   around which droplets of water formed.

3. In some cases, small silver iodide crystals are dropped into clouds to seed them and make rain. What do you think the crystals do?

   Like the particles of smoke, the crystals serve as condensation nuclei to help

   form water droplets.

4. Based on your results, does the water temperature or the presence of smoke have a greater effect on cloud formation?

   Answers may vary. Accept all reasonable responses as long as they seem to

   correlate with the students' experimental results.

HRW material copyrighted under notice appearing earlier in this work.

Unit 7: Performance-Based Test

5. Write a hypothesis that explains how squeezing the bottle caused clouds to form.

Answers may vary. Accept all reasonable hypotheses. A sample response might
be that squeezing the bottle raised the pressure, which also raised the air
temperature inside the bottle. This increased the amount of water vapor in the
air. Releasing the bottle decreased the pressure inside the bottle and lowered
the air temperature. The cooler air could not hold as much moisture as it could
when it was warmer, so the water vapor condensed into clouds.

_____

_____

6. Explain your results in terms of relative humidity and dew point.

Both the water temperature and the number of squeezes determined the relative
humidity of the air in the bottle. The warmer water required fewer squeezes to
raise the relative humidity of the air to nearly its saturation point. Releasing
the bottle allowed the air in the bottle to cool and clouds to form as the air
temperature reached the dew point.

_____

7. Estimate the probable amount of cloudiness and rainfall for the following climates. Rank the climates from most cloudy/rainy (1) to least cloudy/rainy (6).

_____ cool, moist, not dusty   _____ hot, dry, dusty

_____ cool, dry, dusty   _____ warm, moist, not dusty

_____ hot, moist, dusty   _____ warm, moist, dusty

Explain your ranking based on the results of your experiment.

Answers may vary. The rankings will depend on whether students place
importance on condensation nuclei or the temperature in cloud formation.
The hot, moist, and dusty climate should rank first, and one of the cool climates
should rank last.

_____

_____

_____

_____

_____

M O D E R N   E A R T H   S C I E N C E

Unit 7: Performance-Based Test

8. In which of the climates listed in Question 7 do you think cloud seeding would be most effective? least effective?

   Answers may vary. Accept all reasonable responses that are well supported.

   A typical answer would be that cloud seeding is most effective in a warm,

   moist, dustless climate and least effective in a cool, dry, dusty climate.

9. What might be done in the activity to improve the accuracy of data and the conclusions taken from the data?

   Answers will vary but should suggest reducing the number of uncontrolled

   variables or minimizing experimental error through repeated trials. Typical

   answers may include repeating trials of each step, measuring the water

   temperature, using a wider range of water temperatures, and controlling the

   amount of smoke for each run.

MODERN EARTH SCIENCE

Unit 8: Studying Space

Performance-Based Test:
Putting Planets in Motion

Objectives
Investigate planetary motion by testing Kepler's laws. Use Kepler's laws to solve several hypothetical problems.

Background
In 1600, a young German named Johannes Kepler arrived in Prague. He quickly became the protégé of Tycho Brahe, the leading astronomer of the time. Brahe's observations of Mars moving through the sky formed the basis for Kepler's work. Over the next several decades, Kepler proposed three laws that he believed governed the motions of all planets and moons.

Kepler's first law states that planets move around the sun in elliptical orbits, with the center of the sun located at one focus of the ellipse.

The second law states that an imaginary line drawn from the sun to a planet sweeps out equal areas in equal times (See Figure 1 below.). This means that planets move faster when they are closer to the sun.

The third law states that the time a planet takes to complete an orbit—its orbit period—is proportional to the radius of the orbit. More specifically, the period of rotation squared is proportional to the orbit's radius cubed, or $p^2 = k(r^3)$.

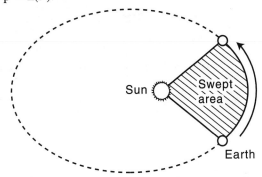

Figure 1. Area swept by an imaginary line

Before You Begin
Read the following guidelines for completing this test.
- Keep in mind that your teacher will be observing and grading your in-class behavior as well as your written responses. In particular, your teacher will be noting your ability to follow the given procedures, your adherence to classroom or laboratory safety, and your methods and reasoning in solving the problems.
- Try not to let what others are doing influence your work. Remember that a problem often has several acceptable solutions.
- Do not talk to other students unless you are working in a group. Talk only to members of your group and try not to disturb other students.
- Use only the materials provided.

Safety Alert
Scissors are sharp objects and should be handled carefully. Make sure that your group completes this activity at a safe distance from the other groups. Wear safety goggles.

M O D E R N   E A R T H   S C I E N C E

Unit 8: Performance-Based Test

Procedure

Step 1: Tie a loop in one end of the string and thread it through the spool, as shown in Figure 2. Tie several washers to the other end. Hook the spring scale onto the loop, and adjust the string until the washers are 1 m from the spool. Mark the string at this point.

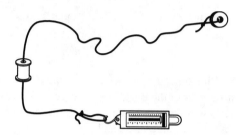

Figure 2. Experimental setup

Step 2: Hold the spool and the scale, and swing the washers in a circle just fast enough to keep the string taut (Figure 3). Practice swinging the washers until you can maintain a constant force on the scale. Record the amount of force in the data table on the next page. Continue to swing the washers while your group members do Step 3.

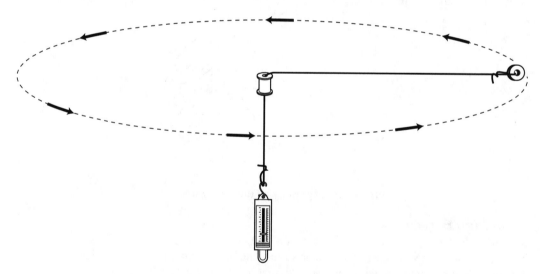

Figure 3. Swinging the weights

Step 3: Using a stopwatch, measure and record the time it takes for the washers to complete ten complete orbits. Record your results in the data table. Also calculate and record the period, or the time it takes to complete one orbit.

Step 4: Shorten the string by 15 cm and repeat Steps 2 and 3. Make sure that the force on the scale is the same as before. Record your data in the table. Repeat this step three more times, shortening the string by 15 cm each time.

M O D E R N   E A R T H   S C I E N C E

Unit 8: Performance-Based Test

Step 5: Let out the string to its original length. While swinging the washers, pull down and let up on the scale to create an elliptical orbit. Carefully observe the force on the string and the relative speed of the washers. Record your observations in the space provided below.

Students should observe that the force applied to the washers does not vary

through the orbit. Students should also notice that the washers move faster

when they are closer to the spool.

_____

_____

Data Table

A	B	C	D	E
String length (radius), or r (cm)	Time/10 orbits (sec.)	Orbit period, or p (sec.)	p^2/r^3	Force (N)
Data will vary, depending on the force the students use when swinging the washers. Data quality should be reflected by consistency in the values in column D. The closer the values are for each trial, the better the experimental procedure.				

Questions

Answer the following questions in the space provided. Support your answers by explaining your reasoning.

1. What do the washers, the spool, and the string represent in this experiment?

   The washers represent an orbiting moon or planet. The spool represents a planet

   or star that the washers are orbiting. The string represents the force of gravity

   holding the moon or planet in orbit.

   _____

   _____

2. Use the space provided on page 59 to plot a graph of the data in columns C and D.
3. Kepler's third law states that the orbit period of a moon or planet is related to the average radius of its orbit. Do your results support this idea?

   Students should realize that a nearly constant value of the ratio in column D

   supports Kepler's third law.

   _____

   _____

   _____

Unit 8: Performance-Based Test

4. Figure 4 (below) illustrates the elliptical orbit that you simulated in Step 5. On the diagram, mark the point at which the washers were moving the fastest with an F. Mark the point at which the washers were moving the slowest with an S.

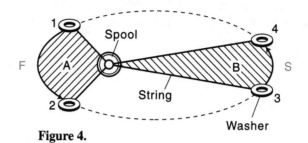

Figure 4.

5. In Figure 4, the areas swept by the string are equal (area of *A* = area of *B*). Based on this fact and your observations from Step 5, what can you infer about the time it takes the washers to go from *1* to *2* and from *3* to *4?* Explain your answer using Kepler's laws.

According to Kepler's third law, the time it takes the washers to go from *1* to *2*

must be equal to the time it takes to go from *3* to *4.* Given that the lengths of the

arcs are not equal but that the areas *A* and *B* are equal, the washers must move

faster the closer they are to the spool. This is consistent with Kepler's second

law and should be consistent with students' observations in Step 5.

_____

_____

_____

6. Chasing and docking with another spacecraft, such as the shuttle docking with Russia's Mir space station, is very tricky. Using Kepler's laws, explain why such maneuvers are difficult.

Answers may vary. Accept all reasonable responses. A possible answer is that

according to Kepler's third law, any change in speed may potentially change the

shuttle's orbital radius, unless its trajectory is carefully controlled.

_____

_____

_____

_____

MODERN EARTH SCIENCE

Unit 8: Performance-Based Test

Graphs will vary, depending on the data, but should meet the following criteria:
- The vertical axis is scaled in seconds.
- The horizontal axis is scaled in centimeters.
- Two curves are drawn, one for period and one for the ratio in column *D*.
- The two curves should be continuous and labeled appropriately.
- The ratio curve should approximate a straight line.
- There should be labels, a title, and a key to explain the graph.